T0137116

Springer Theses

Recognizing Outstanding Ph.D. Research

Aims and Scope

The series "Springer Theses" brings together a selection of the very best Ph.D. theses from around the world and across the physical sciences. Nominated and endorsed by two recognized specialists, each published volume has been selected for its scientific excellence and the high impact of its contents for the pertinent field of research. For greater accessibility to non-specialists, the published versions include an extended introduction, as well as a foreword by the student's supervisor explaining the special relevance of the work for the field. As a whole, the series will provide a valuable resource both for newcomers to the research fields described, and for other scientists seeking detailed background information on special questions. Finally, it provides an accredited documentation of the valuable contributions made by today's younger generation of scientists.

Theses are accepted into the series by invited nomination only and must fulfill all of the following criteria

- They must be written in good English.
- The topic should fall within the confines of Chemistry, Physics, Earth Sciences, Engineering and related interdisciplinary fields such as Materials, Nanoscience, Chemical Engineering, Complex Systems and Biophysics.
- The work reported in the thesis must represent a significant scientific advance.
- If the thesis includes previously published material, permission to reproduce this must be gained from the respective copyright holder.
- They must have been examined and passed during the 12 months prior to nomination.
- Each thesis should include a foreword by the supervisor outlining the significance of its content.
- The theses should have a clearly defined structure including an introduction accessible to scientists not expert in that particular field.

More information about this series at http://www.springer.com/series/8790

Adam Robert Vernon

Collinear Resonance Ionization Spectroscopy of Neutron-Rich Indium Isotopes

Doctoral Thesis accepted by
The University of Manchester, Manchester,
United Kingdom

 Springer

Author
Dr. Adam Robert Vernon
Instituut voor Kern- en Stralingsfysica
KU Leuven
Heverlee, Belgium

Supervisor
Prof. Kieran Flanagan
Department of Physics and Astronomy
The University of Manchester
Manchester, UK

ISSN 2190-5053   ISSN 2190-5061 (electronic)
Springer Theses
ISBN 978-3-030-54191-0   ISBN 978-3-030-54189-7 (eBook)
https://doi.org/10.1007/978-3-030-54189-7

This Springer imprint is published by the registered company Springer Nature Switzerland AG
The registered company address is: Gewerbestrasse 11, 6330 Cham, Switzerland

Supervisor's Foreword

Nuclear physics is currently experiencing an exciting period of rapid gains in both theory and experimental tools. Our understanding of the complex many-body nuclear system and the emergence of simple structural properties has accelerated significantly in the last 10 years. This is not just a result of enhanced computational power but also the development of elegant physics and algorithmic methods that have established a paradigm. This has enabled ab-initio calculations based on assumptions associated with the lightest nuclear systems to be performed on nuclei as heavy as ^{100}Sn. Experimental techniques have been refined to enable more exotic nuclear system to be studied and provide robust verification of theoretical predictions. Laser spectroscopy is one such technique that can measure nuclear observables without introducing assumptions associated with any particular nuclear model. The most stringent test of the latest nuclear models requires measurements to be made far from stability on isotopes that can only be produced at high energy accelerator laboratories. The most interesting isotopes are typically produced at the lowest rates with the highest levels of contamination. This has demanded continuous development of laser spectroscopy techniques.

Adam Vernon's Ph.D. project focused on the measurement of nuclear moments and charge radii in neutron-rich indium isotopes. He has spent a significant part of his thesis work understanding the atomic charge-exchange process and developing code to calculate the population of atomic states after the exchange of an electron. This allowed efficient resonance ionization schemes to be developed. He then contributed to the development of an ablation ion source that allowed experimental validation and model comparison. This was critical for performing nuclear measurements on exotic radioactive atoms. He performed the first measurements on neutron-rich indium isotopes. After careful analysis of the spectroscopic data collected at CERN, he observed a dramatic change in structure as the $N = 82$ shell closure is approached.

Manchester, UK   Prof. Kieran Flanagan

Abstract

This thesis presents laser spectroscopy measurements of the hyperfine structures of the neutron-rich indium isotopes $^{113-131}$In, which allowed the determination of changes in root-mean-square nuclear charge radii, nuclear spins, magnetic dipole moments and quadrupole moments. These measurements were made at CERN-ISOLDE using the Collinear Resonance Ionization Spectroscopy (CRIS) setup, a high-sensitivity, high-resolution technique.

The development of a laser ablation ion source setup allowed the investigation of multi-step laser ionization schemes and measurements of the hyperfine parameters of several atomic states in indium. This lead to the choice of the 246.8 nm (5p ^{2}P$_{3/2}$→ 9s ^{2}S$_{1/2}$) and 246.0 nm (5p ^{2}P$_{1/2}$→ 8s ^{2}S$_{1/2}$) atomic transitions as the most appropriate for the extraction of the nuclear observables. The relative atomic populations of the 5p ^{2}P$_{3/2}$ and 5p ^{2}P$_{1/2}$ states were measured using the laser ablation ion source and were found to be consistent with relative atomic population simulations.

The newly measured hyperfine parameters of the 8s ^{2}S$_{1/2}$ and 9s ^{2}S$_{1/2}$ states were compared to relativistic coupled-cluster calculations of atomic structure parameters, which gave confidence in the calculated values and enabled the nuclear quadrupole moments in indium to be extracted with unprecedented accuracy.

Further comparison to specific mass shift and field shift values extracted from the isotope shift measurements aided in the development of an 'analytic response' approach to determine isotope shift constants. This allowed for the first isotone independent determination of the nuclear charge radii of the indium isotopes and of the absolute charge radii for an odd-proton system near the $Z = 50$ shell closure.

The nuclear-model-independent ground and isomeric state electromagnetic properties of indium measured up to $N = 82$ in this work, particularly of the $I = 9/2^+$ and $I = 1/2^-$ states will refine the understanding of the nuclear structure in this region of the Segrè chart. A sudden increase in the $I = 9/2^+$ state μ values, deviation from the Schmidt value of the $I = 1/2^-$ state μ values and disappearance of odd-even staggering in the ground state mean-square charge radii, among over observations, are yet to be explained by nuclear theory.

Acknowledgements

I would like to express my gratitude to my supervisor Kieran Flanagan for giving me the opportunity to do this work and my co-supervisor Ronald Garcia Ruiz for helping me find the unique opportunities that can be found in unexpected results. They have been constant the source of enthusiasm over the last 3 years, and I am greatly indebted to both.

I would like to acknowledge everyone in the extended CRIS team for their dedication to the experiment, which has made it a productive but also enjoyable environment: Jonathan Billowes, Cory Binnersley, Mark Bissell, Thomas Cocolios, Greg Farooq-Smith, Wouter Gins, Ruben de Groote, Ágota Koszorús, Gerda Neyens, Kara Lynch, Fred Parnefjord-Gustafsson, Chris Ricketts, Shane Wilkins and Xiaofei Yang. Their contributions over the years have been invaluable towards making this work possible. And thanks to Shane for taking the time to proof-read several chapters of this thesis, your suggestions were much appreciated.

I would like to thank Bijaya Sahoo for developing atomic calculations which greatly enhanced the value of the results of the work described in this thesis.

I would also like to thank everyone at ISOLDE for creating such a hospitable community during my time here and for their part in enabling the experiments over the last 2 years to be successful.

Finally, I would also like to thank my family for their support over the past few years.

Thanks to all of you, and anyone I forgot to acknowledge!

February 2019   Adam Robert Vernon

Contents

Chapter 1
Introduction

1.1 Introduction

The nuclear force, when combined with the Pauli principle in the finite many-body system of atomic nuclei cause remarkably simple, but yet often unpredictable, trends to emerge. It is expected that the microscopic description of the nucleus is completely consistent with the collective pictures of deformation and volume, that we are only held back by sufficient computational power [1, 2]. Surprisingly simple dependences on proton or neutron number appear for even heavy systems [3–8]. Explanation of these simple patterns from first principles is a main focus for modern nuclear theories [9–11].

The measurement of nuclear observables over isotope chains near doubly-magic nuclei give an insight into the evolution of single-particle and collective properties of the nucleus and their interplay. The electromagnetic properties of the nucleus measured by laser spectroscopy allow changes in both valance-proton orbitals and collective properties of the nucleus to be probed in a nuclear-model independent way.

With a proton hole in the $Z = 50$ shell closure, the neutron-rich indium isotopic chain ($Z = 49$) offers a compelling scenario to explore the evolution of nuclear-structure properties towards the doubly-magic isotope ^{132}Sn ($N = 82$). The hole in the $\pi 1g_{9/2}$ orbit creates a chain of $I = {}^9\!/_2{}^+$ nuclear ground states between $N = 50$ and $N = 82$ for the odd mass indium isotopes. A proton from the $\pi 2p_{1/2}$ orbit can be excited (<400 keV) to the $\pi 1g_{9/2}$ orbit to create a $\pi 2p_{1/2}^{-1}$ configuration, which creates an accompanying chain of long-lived $I = {}^1\!/_2{}^-$ isomer states.

The neutrons of the even mass indium isotopes occupy the $\nu 3s_{1/2}$, $\nu 2d_{3/2}$ and $\nu 1h_{11/2}$ orbitals, allowing a variety of long-lived and high-spin configurations to exist when coupled with the $\pi 1g_{9/2}$ or $\pi 2p_{1/2}$ proton states. This gives rise to multiple isomeric states at each isotope mass, making measurements of the indium isotope chain a very efficient probe of evolution of nuclear structure in this region of the nuclear chart and its dependence on the particle or hole nature of the proton.

© The Editor(s) (if applicable) and The Author(s), under exclusive license
to Springer Nature Switzerland AG 2020
A. R. Vernon, *Collinear Resonance Ionization Spectroscopy
of Neutron-Rich Indium Isotopes*, Springer Theses,
https://doi.org/10.1007/978-3-030-54189-7_1

The $I = \frac{9}{2}^+$ ground states of $^{103-127}$In are a textbook example [12] of single-particle characteristics persisting with varying neutron number. The magnetic dipole moments of the $I = \frac{9}{2}^+$ states were measured to have a remarkably constant value of $\mu = 5.43$ to 5.53 μ_N over a range of 22 isotopes [13, 14]. At the same time, a yet unexplained decrease across the Schmidt line of the magnetic moments for the $I = \frac{1}{2}^-$ states with neutron number was observed [14]. These deviations are often explained by first-order configuration mixing, however for odd protons in the $j = l - \frac{1}{2}$ orbit, first-order configuration mixing is not expected to be present [15], pointing to higher-order contributions as an explanation. The new measurements of these $I = \frac{1}{2}^-$ states and the other long-lived (>200 ms) isomer states of indium presented in this work will contribute to a better understanding of the nuclear structure in the $Z = 50$ region.

The primary outcome of the work described in this thesis was the measurement of the nuclear ground and long-lived isomeric state electromagnetic properties of the exotic indium isotopes $^{113-131}$In. This includes the nuclear spin, nuclear magnetic dipole moment, nuclear electric quadrupole moment and changes in mean-squared charge radii. These measurements were enabled and guided by theoretical and technical developments which are described in this thesis in the following sequence.

An introduction to the theory needed to connect laser spectroscopy measurements to the extraction of nuclear observables is given in Chap. 2, in addition to the nuclear structure background used to interpret the measurements of the indium isotopes performed in this work. As the measurements were made on the atoms rather than ions of the isotopes, the ion beam first needed to be neutralised by atomic charge exchange, details of the neutralisation process and of calculations made to predict the relative atomic populations of the indium isotopes following neutralisation are outlined in Sect. 2.4.

Then in Chap. 3 an overview is given of the experimental methods used to produce the exotic indium isotopes at CERN-ISOLDE, and prepare them for high-resolution laser spectroscopy measurements at the Collinear Resonance Ionization Spectroscopy (CRIS) beamline.

A description of the required laser spectroscopy setup which was shared by all experiments described in this thesis is given in Chap. 4. The various multi-step laser ionization schemes which were explored are discussed in this chapter. A technical summary of improvements made to the CRIS setup during this time is also included in the paper:

I. **A.R. Vernon**, R.P. de Groote et al. **Optimising the Collinear Resonance Ionization Spectroscopy experiment**, Nucl. Instruments Methods Phys. Res. Sect. B Beam Interact. with Mater. Atoms, doi:10.1016/j.nimb.2019.04.0495

In Chap. 5 the development of a laser ablation ion source setup for high-resolution laser spectroscopy is described. A description of this work, and the atomic physics probed by the measured hyperfine transitions of the naturally occurring indium isotopes ^{113,115}In are also contained in a briefer format in the paper:

II. R.F. Garcia Ruiz, **A.R. Vernon**, C.L. Binnersley, B.K. Sahoo et al. **High-Precision Multiphoton Ionization of Accelerated Laser-Ablated Species**, Phys. Rev. X. 8 (2018) 041005. doi:10.1103/PhysRevX.8.041005 - **Featured viewpoint article in APS physics**.

This work led to the choice of the atomic transitions later used for the measurements of the neutron-rich indium isotopes $^{113-131}$In, in order to extract their nuclear structure observables.

Atomic parameter calculations [16] of magnetic dipole hyperfine structure constants were tested by comparison with the measurements of ^{113,115}In made in this work. This led to improved calculations [16] of the electric field gradient factors, later used to extract the nuclear electric quadrupole moments with improved accuracy for the neutron-rich $^{113-131}$In isotope measurements. The laser ablation ion source setup also allowed measurement of the relative atomic populations of indium, which are discussed and compared to the calculations in Sect. 5.9. These results and further neutralisation calculations for elements $1 \leq Z \leq 89$ are also contained in the paper:

III. **A. R. Vernon** et al., **Simulations of atomic populations of ions $1 \leq Z \leq 89$ following charge exchange tested with collinear resonance ionization spectroscopy of indium**, Spectrochim. Acta Part B At. Spectrosc., vol. 153, pp. 61–83, Mar. doi:10.1016/j.sab.2019.02.001

The analysis performed to extract of the electromagnetic properties of the ground and isomeric states from the measurements of the $^{113-131}$In isotopes is described in Chap. 6.

Accurate atomic-field-shift and specific mass shift parameters are needed to evaluate the nuclear mean-square charge radii from isotope shift measurements. Experimental evaluation of the atomic factors by comparison to absolute charge radii measurements has limited accuracy for indium, as it has only two naturally occurring isotopes ^{113,115}In. Therefore, in collaboration with atomic theorists (B. K. Sahoo), calculations were performed [17] using a relativistic coupled-cluster method to determine these atomic factors. The calculations were compared to field-shift and specific-mass-shift values extracted from analysis of the measured isotope shift values of indium. This process is described in Sect. 6.5.1 of Chap. 6. The calculated mass shift and field shift factors allowed evaluation of the nuclear mean-square charge radii of the indium isotopes independent of its neighbouring isotones, with a high accuracy compared to competing approaches. As the case is exemplary of all odd-Z nuclei, and due to the accuracy of the calculation method, the results of this section were also submitted in the paper:

IV. B. K. Sahoo, **A. R. Vernon** et al., **Analytic Response Relativistic Coupled-Cluster Theory: The first application to In isotope shifts**, New J. Phys. 22, 012001 (2020) doi:10.1088/1367-2630/ab66dd

In Chap. 7 an interpretation of these results within the context of nuclear structure in the $Z = 50$ region and in general is given. Finally, conclusions are made in Chap. 8.

References

1. Launey KD, et al (2017) Bulg J Phys 44:345. http://www.bjp-bg.com/papers/bjp2017_4_345-356.pdf
2. Freer M, et al (2018) Rev Mod Phys 90(3):035004. ISSN 0034-6861. https://link.aps.org/doi/10.1103/RevModPhys.90.035004
3. Marsh BA, et al (2018) Nat Phys 1. ISSN 1745-2473. http://www.nature.com/articles/s41567-018-0292-8
4. Garcia Ruiz RF et al (2016) Nat Phys 12(6):594. ISSN 1745-2473. http://www.nature.com/articles/nphys3645 https://doi.org/10.1038/nphys3645
5. Yordanov DT, et al (2016) Phys Rev Lett 116(3):1. ISSN 10797114
6. Hakala J, et al (2012) Phys Rev Lett 109(3):032501. ISSN 0031-9007. https://doi.org/10.1103/PhysRevLett.109.032501
7. Zawischa D (1985) Phys Lett B, 155(5–6):309. ISSN 0370-2693. https://www.sciencedirect.com/science/article/pii/037026938591576X
8. Wilkinson D (1977) Nucl Instrum Methods 146(1):143. ISSN 0029554X. http://linkinghub.elsevier.com/retrieve/pii/0029554X77905092
9. Hagen G, et al (2014) Rep Prog Phys 77(9):096302. ISSN 00344885. http://arxiv.org/abs/1312.7872 http://stacks.iop.org/0034-4885/77/i=9/a=096302?key=crossref.e42bdb60fa516e13738c4a5f1bb833f1
10. Sun ZH, et al (2018) Phys Rev C 98(5):054320. ISSN 2469-9985. https://doi.org/10.1103/PhysRevC.98.054320
11. Hergert H, et al (2016) Phys Rep 621:165. ISSN 0370-1573. https://www.sciencedirect.com/science/article/pii/S0370157315005414?via%3Dihub
12. Heyde KLG (1990) The nuclear shell model. Springer series in nuclear and particle physics. Springer, Berlin, Heidelberg. https://doi.org/10.1007/978-3-642-97203-4 https://books.google.ch/books?id=aBz4CAAAQBAJ&dq=heyde+nuclear+shell+model&source=gbs_navlinks_s
13. Stone N (2005) Atom Data Nucl Data Tables 90(1):75. ISSN 0092640X. http://linkinghub.elsevier.com/retrieve/pii/S0092640X05000239
14. Eberz J, et al (1987) Nucl Phys A 464(1):9. ISSN 03759474. http://linkinghub.elsevier.com/retrieve/pii/0375947487904192
15. Arima A, et al (1954) Prog Theor Phys 12(5):623–641. ISSN 0033-068X. https://academic.oup.com/ptp/article-lookup/doi/10.1143/PTP.12.623
16. Garcia Ruiz RF, et al (2018) Phys Rev X 8(4):041005. ISSN 160-3308. https://doi.org/10.1103/PhysRevX.8.041005
17. Sahoo BK, et al (2020) New J Phys 22(1):012001. ISSN 1367-2630. https://iopscience.iop.org/article/10.1088/1367-2630/ab66dd

Chapter 2
Background Atomic and Nuclear Physics

This chapter gives a brief summary of the underlying principles of atomic and nuclear physics which are needed to understand the laser spectroscopy methods used to measure the short-lived indium isotopes studied in this thesis, in addition to their relevance to nuclear structure.

2.1 Nuclear Structure

The nuclear many-body problem can be written succinctly by the Schrödinger equation, with the Hamiltonian, H, as

$$H \, |\psi\rangle = (T + V) \, |\psi\rangle = E \, |\psi\rangle \tag{2.1}$$

where T is the kinetic energy operator and V is the potential energy operator, for a system of A nucleons [1].

$$V = \sum_{i<j} v_{ij} + \sum_{i<j<k} v_{ijk} + \sum_{i<j<k<l} v_{ijkl} + \dots \tag{2.2}$$

where the number of subscript indices stands for the number of bodies in the interaction, e.g. v_{ijk} denotes 3-body interactions. To make the problem more tractable, calculation of orders higher than two-body are often neglected. However three-body interactions have been shown to be essential for nuclear structure, in calculations of lighter systems [2–4] where the problem is still computationally manageable.

The Hamiltonian, H, can be re-written as [1]

© The Editor(s) (if applicable) and The Author(s), under exclusive license
to Springer Nature Switzerland AG 2020
A. R. Vernon, *Collinear Resonance Ionization Spectroscopy
of Neutron-Rich Indium Isotopes*, Springer Theses,
https://doi.org/10.1007/978-3-030-54189-7_2

$$H = T + \sum_i U_i + V - \sum_i U_i$$

$$= H_0 + H_{res}$$

(2.3)

by defining a mean-field potential, U_i, where H_0 then describes independent nucleon motion with the many-body interactions as a residual correction, H_{res}.

2.1.1 The Nuclear Shell Model

In the independent-particle shell model, a mean-field potential is chosen to describe the independent motion of a nucleon in presence of all the other nucleons in the nucleus generating that potential i.e. $H = H_0$.

A simple form of the mean-field potential is a harmonic-oscillator potential [5], with additional spin-orbit $\mathbf{l}.\mathbf{s}$ and orbital angular momentum squared, l^2, terms[1]

$$U(r) = \frac{1}{2}m\omega^2 r^2 + C\mathbf{l}.\mathbf{s} + Dl^2 \ .$$

(2.4)

The harmonic-oscillator potential alone reproduces the magic number energy gaps 2, 8, 20, which are experimentally observed in the single-particle energies of light stable isotopes. It was found by [7] that the inclusion of the $\mathbf{l}.\mathbf{s}$ and l^2 terms were necessary to reproduce the higher magic numbers 28, 50 ... which were observed for heavier naturally abundant nuclei.

The shell-model occupancies for the neutron-rich indium isotopes $^{113-131}$In studied in this thesis are shown in Fig. 2.1. Where the single-particle energies are indicated for the simple harmonic oscillator, and for the addition of the l^2 and $\mathbf{l}.\mathbf{s}$ terms. The protons and neutrons have their own unique quantum number and therefore their orbital occupancies are considered separately in the shell model.

Two- or higher-body interactions are often needed to reproduce experimental values [8], and so interacting shell models are largely used today. However this simplified model still has use as a guide in assessing the types of interactions [9] present in a system and the order of the many-body calculations expected to be needed to reproduce experiment. As the unpaired nucleons determine the final spin of the nucleus, the shell model orbitals can be used to predict the ground-state spin of many nuclei [10]. See Fig. 2.3 for the ground-state spins of $^{113-131}$In.

---

[1]A Woods-Saxon [6] potential with the additional spin-orbit $\mathbf{l}.\mathbf{s}$ interaction, is also a common potential as it allows for fitting to known nuclei and surface diffuseness.

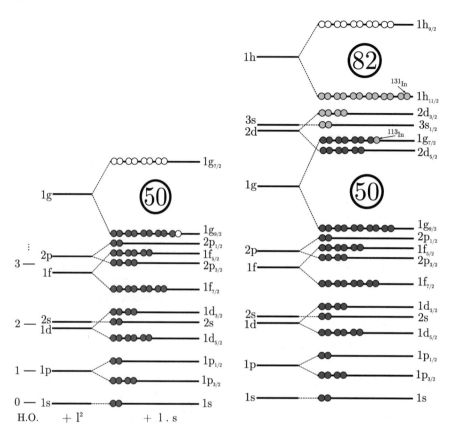

Fig. 2.1 The leftmost lines indicate the levels created by quantised angular momentum in a simple potential well, the levels to the right include the splitting from the spin-orbit effect and show the basic form of the shell model. The circles indicate the multiplicity of these states. Filled red circles indicate the proton occupancy of the ground state in indium isotopes. While the blue-filled circles indicate the neutron occupancy for up to ^{113}In, green-filled circles indicate the range of occupancies for masses $^{113-131}$In

2.1.2 Ground State Observables

Similar to the potential energy in Eq. 2.2, the electromagnetic charge density $\rho_c(r)$ and current density $j(r)$ can also be written [11] as a sum of its many-body operators as

$$\rho_c(r) = \sum_{i<j} \rho_c(r)_{ij} + \sum_{i<j<k} \rho_c(r)_{ijk} + \sum_{i<j<k<l} \rho_c(r)_{ijkl} + \dots , \qquad (2.5)$$

and

$$j(r) = \sum_{i<j} j(r)_{ij} + \sum_{i<j<k} j(r)_{ijk} + \sum_{i<j<k<l} j(r)_{ijkl} + \dots , \qquad (2.6)$$

respectively. If the charge density and currents in the nucleus were known then the electromagnetic properties of the nucleus would also be fully described. However these quantities are not directly observable. The impulse approximation [11] is often used to construct non-relativistic single-nucleon operators. Under this approximation the unmodified experimental g-factors of the free neutrons and protons are used in one-body operators applied to finite nuclei. Therefore terms which require two-body or higher order currents are neglected, such as meson-exchange currents.

Under this approximation and a multi-pole expansion of the magnetic and electric fields in the nucleus, the following electromagnetic operators dominate the contributions observable in hyperfine structures accessible with the present precision of modern laser spectroscopy measurements of exotic isotopes: the nuclear mean-square charge radius $\langle r^2 \rangle$, the nuclear magnetic dipole moment μ and the nuclear electric quadrupole moment Q_S. Higher order moments such as the magnetic octupole moment have also been measured with laser spectroscopy, in cases where they are accessible in certain heavier systems [12, 13] or with dedicated techniques [14, 15].

2.1.2.1 Nuclear Mean-Squared Charge-Radius

The mean-square charge-radii give a chiefly macroscopic view of the nucleus, providing information on changes in its shape, volume and surface diffuseness. The nuclear charge radii of exotic isotopes are one of the most sensitive nuclear observables available for testing many-body calculations and the description of the nuclear force [16–18]. For a charge-density distribution $\rho(\vec{r})$ the definition of the mean-square charge-radius is given by

$$\langle r^2 \rangle = \frac{\int_0^\infty \rho(\vec{r}) r^2 \mathrm{d}^3 r}{\int_0^\infty \rho(\vec{r}) \mathrm{d}^3 r} . \qquad (2.7)$$

In the case of spherical symmetry and a homogeneous distribution then this can be approximated by the spherical droplet-model value [19]

$$\langle r^2 \rangle_0 = \frac{3}{5} r_0^2 A^{2/3} , \qquad (2.8)$$

where the empirical parameter $r_0 = 1.18$ fm [20] is often used, and A is the atomic number. In the case of charge distribution with deformation the spherical harmonics [21] can be included as

$$\langle r^2 \rangle = \langle r^2 \rangle_0 \left(1 + \frac{5}{4\pi} \sum_{\lambda=2}^{\infty} \langle \beta_\lambda^2 \rangle \right) , \qquad (2.9)$$

where $\langle r^2 \rangle_0$ is an equivalent volume spherical nucleus and $\langle \beta_\lambda^2 \rangle$ are the deformation parameters of order λ [22]. In the first-order approximation [23] this is given by

$$\lim_{\lambda \to 2} \langle r^2 \rangle = \langle r^2 \rangle_0 \left(1 + \frac{5}{4\pi} \langle \beta_2^2 \rangle \right) , \qquad (2.10)$$

See Fig. 2.2 for a diagram indicating the β_2 parameter in the strong-coupling limit.
The $\langle r^2 \rangle$ are not direct observables from laser spectroscopy experiments however, instead changes in $\langle r^2 \rangle$ between isotopes are measured, the difference is therefore given by

$$\delta \langle r^2 \rangle^{A,A'} = \delta \langle r^2 \rangle_0^{A,A'} \left(1 + \frac{5}{4\pi} \sum_\lambda \delta \langle \beta_\lambda^2 \rangle^{A,A'} \right) , \qquad (2.11)$$

for the change between isotopes of masses A and A'. Where $\lambda = 2$ is the quadrupole deformation parameter, $\lambda = 3$, the octupole deformation parameter etc. Hence simultaneous changes in deformation and volume are measured by $\delta \langle r^2 \rangle^{A,A'}$. As the $\langle \beta_2^2 \rangle$ term is the square of the deformation parameter the sign of the deformation is therefore not inferable from $\langle r^2 \rangle$ alone.

2.1.2.2 Nuclear Magnetic Dipole Moment and Nuclear Spin

The nuclear spin, I, and resulting nuclear magnetic dipole moment, μ, give a more microscopic view of the motion of the nucleons. For many nuclei the relationship

$$\mu = gI\mu_N , \qquad (2.12)$$

is used [24] under the impulse approximation. Where the nuclear magneton has the value $\mu_N = \frac{e\hbar}{2m_p} = 5.050783699(31) \times 10^{-27}$ J/T [25], and g is the gyromagnetic factor which then contains the nuclear structure information besides the spin of nucleus I. The pairing force [26] gives a strong binding increase for coupling nucleon pairs to total spin $I = 0$. The total spin of the nucleus is then determined by unpaired valance nucleons in the proton and neutron shells.
The magnetic dipole moment operator separated into contributions from its orbital and spin components of the proton (π) and neutron (ν) orbital motion, given [24] by

$$\mu = \left(\sum_{i=0}^{Z} g_L^\pi \mathbf{L}_i + \sum_{i=0}^{Z} g_S^\pi \mathbf{S}_i + \sum_{j=0}^{N} g_L^\nu \mathbf{L}_j + \sum_{j=0}^{N} g_S^\nu \mathbf{S}_j \right) , \qquad (2.13)$$

where the orbital and spin g-factors have their free-particle values, $g_L^\pi = 1$, $g_L^\nu = 0$, $g_S^\pi = +5.585694702(17)$ and $g_S^\nu = -3.82608545(90)$ [25]. The magnetic moment is then defined to be the expectation of the z-component of this one-body operator for the magnetic sub-state with maximal spin projection $\mu \equiv \langle I, M = I | \mu_z | I, M = I \rangle$, where $I = 0$ states have no contribution [10]. As mainly unpaired nuclei contribute to μ and their g-factors are expected equal to their single-particle orbit values independent of the total nuclear spin, g-factors are a sensitive probe to changes in the orbital of the valence nucleons.

In the impulse approximation μ_z is a one-body operator acting on individual valence nucleons with total spin I. The μ of a composite state for weak coupling between a valence proton and neutron can be calculated by the additivity rule using jj-type coupling [27, 28],

$$\mu(I) = \frac{I}{2} \left[\frac{\mu(I_p)}{I_p} + \frac{\mu(I_n)}{I_n} + \left(\frac{\mu(I_p)}{I_p} - \frac{\mu(I_n)}{I_n} \right) \frac{I_p(I_p + 1) - I_n(I_n + 1)}{I(I + 1)} \right],$$

(2.14)

for a composite state made from proton spin I_p and neutron spin I_n states, with moments $\mu(I_p)$ and $\mu(I_n)$ respectively. The weak coupling approximation is no longer valid when configuration mixing is likely between the final spin states [27]. For example for the $8^-(\pi 1g_{9/2}^{-1} \otimes \nu 1h_{11/2})$ states of $^{116-128}$In studied in this work, the additivity rule is no longer valid, which is discussed in Chap. 7.

2.1.2.3 Nuclear Electric Quadrupole Moment

A non-spherical nucleus can have an electric quadrupole moment, which is associated with the distribution of protons within the nucleus. This can be measured with a variety of techniques [29] including laser spectroscopy. The electric quadrupole moments provides insight into the occupation of proton orbits, core polarisation and other collective effects (see Sect. 7.3.1). The nuclear electric quadrupole moment operator is defined as

$$\mathbf{Q} \equiv e \sum_{k=1}^{A} (3z_k^2 - r_k^2) \,,$$

(2.15)

where e is the electric charge and z_k, r_k are the position of the kth nucleon [30]. The spectroscopic quadrupole moment, accessible to measurement, is the expectation value of this operator

$$Q_S(I) \equiv \langle I, M = I | \mathbf{Q}_z | I, M = I \rangle \equiv \sqrt{\frac{I(2I - 1)}{(2I + 1)(2I + 3)(I + 1)}} \langle I || \mathbf{Q}_z || I \rangle \,,$$

(2.16)

where the Wigner–Eckhard theorem [31] was used to restate $Q_S(I)$ as a function of the reduced matrix element $\langle I || \mathbf{Q}_z || I \rangle$. It can be seen from this that for states with

$I = \frac{1}{2}$ a spectroscopic quadrupole moment is not defined, however they may still possess an intrinsic quadrupole moment Q_0 [24].

The single-particle quadrupole moment Q_{sp} value can be approximated [24] by

$$Q_{sp} = -e_j \frac{2j - 1}{2j + 2} \langle r_j^2 \rangle ,\tag{2.17}$$

for a nucleon in an orbit j with mean-square charge radius $\langle r_j^2 \rangle$. Where a proton hole has a *universal effective* charge [24] of $-e_j = +e_\pi^{eff} = 1.5e$ [32], giving a prolate deformation to the core.

The primary value of the quadrupole moment is as a measure of the deformation of a nucleus, in the strong-coupling limit of a well defined deformation axis, z, the spectroscopic moment can be related to the intrinsic quadrupole moment by

$$Q_S = \frac{3K^2 - I(I + 1)}{(I + 1)(2I + 3)} Q_0 ,\tag{2.18}$$

where K^2 is the squared projection onto the deformation axis [33] (see Fig. 2.2). A nuclear quadrupole deformation parameter, β, can be defined [10] by a difference in radius between two principle axis of a deformed nucleus ΔR as

$$\beta = \frac{4}{3} \sqrt{\frac{\pi}{5}} \frac{\Delta R}{R_0} ,\tag{2.19}$$

where R_0 is the radius of an equivalent volume spherically symmetric sphere. The radius of a deformed nucleus can then be found by using a second-order spherical harmonic, $Y_{20}(\theta, \phi)$, as

$$R(\theta, \phi) = R_0[1 + \beta Y_{20}(\theta, \phi)] ,\tag{2.20}$$

where θ, ϕ are the spherical co-ordinates. Figure 2.2 shows the spherical-harmonic surface for the prolate ($Q_0 > 0$), spherical ($Q_0 = 0$), and oblate ($Q_0 < 0$) quadrupole moment and nuclear β values.

The intrinsic quadrupole moment, Q_0, can be related [34] to the static quadrupole deformation parameter, β_2^{static}, by

$$\beta_2^{static} = Q_0 \frac{\sqrt{5\pi}}{3Zr_0^2 A^{2/3}} ,\tag{2.21}$$

where r_0 from a parametrization of the droplet model [22, 35] was used in this work. This allows for a measure of the static deformation from the nuclear parameter, β_2^{static}, of the nucleus, albeit with a large nuclear model dependence. Only very few nuclei, those near shell closures, can be considered as spherical ($| \beta | < | 0.1 |$), the majority have prolate or oblate deformation ($| \beta | > | 0.1 |$) [29].

Fig. 2.2 Second-order
spherical harmonic Y_{20}
surface plot showing
spherical, prolate and oblate
nuclear deformation
parameters/quadrupole
moments with respect to the
intrinsic deformation axis z

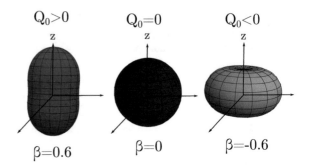

The deformation relationship with mean-square charge radii given by Eq. 2.10 includes both static and dynamic contribution parameters as

$$(\beta_2^{total})^2 = (\beta_2^{static})^2 + (\beta_2^{dyn})^2 , \tag{2.22}$$

when cancelling the static contribution using Eq. 2.21, one can isolate the changes in the dynamic contribution β_2^{dyn} [36].

As $\mathbf{Q}$ is approximately a one-body operator, if the coupling between a valance proton and neutron is weak, then the final $Q_S(I)$ of a nuclear state with spin, I, can be calculated by angular momentum coupling rules [27, 28], which reduces to the additivity formula

$$Q_S(I) = \begin{pmatrix} I & 2 & I \\ -I & 0 & I \end{pmatrix} (-1)^{I_p + I_n + I} (2I + 1)$$

$$\times \left[\begin{Bmatrix} I_p & I & I_n \\ I & I_p & 2 \end{Bmatrix} \frac{Q(I_p)}{\begin{pmatrix} I_p & 2 & I_p \\ -I_p & 0 & I_p \end{pmatrix}} \right.$$

$$\left. + \begin{Bmatrix} I_n & I & I_p \\ I & I_n & 2 \end{Bmatrix} \frac{Q(I_n)}{\begin{pmatrix} I_n & 2 & I_n \\ -I_n & 0 & I_n \end{pmatrix}} \right] \tag{2.23}$$

where $Q_S(I)$ can then be calculated with the proton and neutron quadrupole moments, $Q_S(I_p)$ and $Q_S(I_n)$, in orbits with spins I_p and I_n respectively. The (...) and ... brackets denote the Wigner 3-j and 6-j symbols respectively [27]. The application of this rule to the proton- and neutron-particle states in the even mass indium isotopes is discussed in Chap. 7.

An overview of mean-square charge radii, nuclear magnetic dipole moments, nuclear electric quadrupole moments and nuclear spins of the ground and isomeric states measured in $^{113-131}$In is shown in Fig. 2.3. The observables measured for the first time (to the authors knowledge) in this work are indicated in red, beside the energy level of the states, while observables [37] which were previously measured are shown in black.

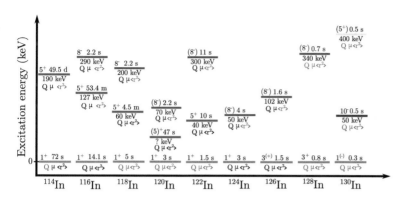

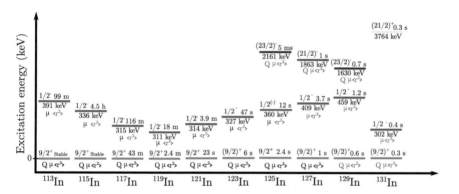

Fig. 2.3 Summary of the ground and isomer states in $^{113-131}$In. New measurements made by CRIS discussed in this work are shown in red under the levels, for the electric quadrupole Q_S, magnetic dipole μ and mean square charge radii $\langle r^2 \rangle$. Those previously measured are shown in black. Data taken from Nuclear Data Sheets [38, 39]. The colours of the energy levels indicate the nuclear states from the same configurations across the isotopes. The excitation energies are not to scale

2.2 Atomic Physics

2.2.1 Hyperfine Structure

The interaction of the magnetic dipole moment of the nucleus, $\vec{\mu}_I$, with the magnetic field generated by its surrounding electron cloud, $\vec{B}_e$, or the interaction of the nuclear electric quadrupole, Q_S, with the electric field gradient of the electrons, $\frac{\partial^2 V}{\partial z^2}$, (where V is the electric potential), are able to break the degeneracy of the fine structure energy levels, $\vec{J} = \vec{L} + \vec{S}$ [23].

Their interaction reveals hyperfine energy levels, F, from the vector coupling of

$$\vec{F} = \vec{I} + \vec{J} , \tag{2.24}$$

for each possible total angular-momentum coupling of the electronic spin, $\vec{J}$ and nuclear spin, $\vec{I}$, the so-called hyperfine structure.

The shift in energy from these two contributions, ΔE_{HFS}, can be expressed [40] as

$$\Delta E_{HFS} = \Delta E_D + \Delta E_Q$$

$$= \overbrace{\frac{C}{2} A_{hf}}^{\alpha} + \overbrace{\frac{3C(C+1) - 4IJ(I+1)(J+1)}{8IJ(2I-1)(2J-1)} B_{hf}}^{\beta} , \tag{2.25}$$

where

$$A_{hf} = \mu \frac{B_e(0)}{\vec{I} \cdot \vec{J}} , \tag{2.26}$$

$$B_{hf} = e Q_S \left\langle \frac{\partial^2 V}{\partial z^2} \right\rangle , \tag{2.27}$$

and

$$C = F(F+1) - J(J+1) - I(I+1) . \tag{2.28}$$

A_{hf} and B_{hf} are the hyperfine structure constants. A_{hf} is the coefficient for the interaction of the magnetic dipole moment, μ, of the nucleus with the magnetic field, $B_e(0)$, at the nucleus.[2] And B_{hf} is the coefficient for the interaction of the nuclear quadrupole moment, Q_S, with the field gradient created by the electrons, $\frac{\partial^2 V}{\partial z^2}$. The convention of spectroscopic quadrupole moments, Q_S, is to use z as the deformation axis, as in Fig. 2.2.

The number of transitions between the hyperfine levels, **F**, is limited by the unit angular momentum of a photon to $\Delta F = 0, \pm 1$, where $F = 0 \rightarrow 0$ transitions are forbidden. An example of these transitions is shown in Fig. 2.4.

Using the approximation that $\left\langle \frac{\partial^2 V}{\partial z^2} \right\rangle$ and $B_e(0)$ remains constant between isotopes of the same element, the ratio of the A_{hf} and B_{hf} factors from Eqs. 2.26 and 2.27 becomes

$$\frac{A_{hf}}{A'_{hf}} \simeq \frac{\mu}{\mu'} \frac{I'}{I} , \tag{2.29}$$

and,

$$\frac{B_{hf}}{B'_{hf}} \simeq \frac{Q_S}{Q'_S} . \tag{2.30}$$

---

[2] $B_e(0)$ and $\frac{\partial^2 V}{\partial z^2}$ are often taken to be constant over the nuclear volume, see Sect. 2.2.2.

Fig. 2.4 Schematic illustration of the hyperfine structure of ^{115}In using the 246.8-nm (5p ^{2}P$_{3/2}$ → 9s ^{2}S$_{1/2}$) transition. Here ω indicates the frequency axis

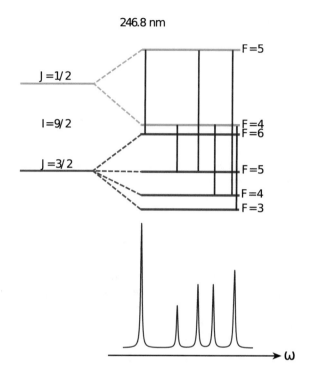

This allows the extraction of a μ value for a given A_{hf} value provided that there is a reference isotope with known μ' and A_{hf} values. These μ values are often determined by NMR measurements with a naturally abundant reference isotope [41].

However for the electric quadrupole moment, it is not possible to apply a sufficiently large and tunable electric field gradient in the laboratory to produce measurable energy differences for measurement of a reference isotope Q_S value. Therefore reported electric quadrupole moment values depend on the calculation of electric field gradients for the atomic environment in which they are measured.

The relative peak intensities of the hyperfine structure can be determined using the line strength, $S_{F_u \rightarrow F_l}$, formula [42]

$$S_{F_u \rightarrow F_l} = \begin{Bmatrix} F_l & F_u & 1 \\ J_l & J_u & I \end{Bmatrix}^2 , \tag{2.31}$$

where {...} is the Wigner 6j-symbol [42, 43]. This relation is valid under the assumption of linear polarised light.

The precision of experiments in an external magnetic field can be extremely high. Measurements of the stable indium isotope ^{115}In allowed determination of the hyperfine octupole constant [14] for the 5p $^2P_{3/2}$ atomic state as $O_{hf}(^{115}\text{In}) = 0.000100(13)$ MHz, compared to $A_{hf}(^{115}\text{In}) = 242.164807(23)$ MHz and $B_{hf}(^{115}\text{In}) = 449.54568(21)$ MHz for the dipole and quadrupole constants. It was not possible to reach this level of precision to determine the octupole moments in this work.

2.2.2 Hyperfine Anomaly

Two additional corrections can be included for phenomena related to the finite size of the nucleus as

$$A_{hf} = A_{hf}^{point}(1 + \epsilon_{BW})(1 + \epsilon_{BR}), \qquad (2.32)$$

where ϵ_{BR} is the Breit-Rosenhal effect contribution due to the extended nuclear charge distribution [44] and ϵ_{BW} is the Bohr–Weisskopf effect contribution from the extended nuclear magnetisation over the nucleus [45, 46]. The A_{hf}^{point} denotes to the A_{hf} factor for a point nucleus in the absence of these effects.

The factors can be introduced as a correction when A_{hf} is determined from ratios, as in Eq. 2.29, giving

$$\frac{A_{hf}}{A_{hf}'} \simeq \frac{\mu}{\mu'}\frac{I'}{I}(1 +^A \Delta^{A'}), \qquad (2.33)$$

where $^A\Delta^{A'}$ is the differential anomaly term. The differential anomaly, $^A\Delta^{A'}$, is a nuclear structure observable in its own right [47], but it is often corrected for or neglected.

The magnitude of this anomaly depends greatly on the isotope studied in addition to the atomic state from which it is measured. The differential hyperfine anomaly measured for the stable ^{113,115}In isotopes were as small as $^A\Delta^{A'} = 0.00075(13)\%$ for the 5p $^2P_{1/2}$ state and $-0.00238(13)\%$ for the 5p $^2P_{3/2}$ state [48, 49]. It was eventually neglected in the following analysis following observation of a constant $\frac{A_u}{A_l}$ for the new isotopes well within experimental uncertainty.

2.2.3 Isotope Shifts

Between isotopes a shift in the electronic energy levels can occur due to differences in mass of the nuclei or field over the nuclei, due to changes in shape or volume. These are termed the mass shift (MS) and field shift (FS) contributions to the isotope shift (IS). They are approximately additive as $IS = MS + FS$. The energy shift is observed as a change in frequency, ν between the same atomic transitions of two isotopes as $\Delta E = h\nu_{A'} - h\nu_A$.

2.2.3.1 Mass Shift

The shift in atomic energy levels between two isotopes due to the electron and isotope masses is termed the 'mass shift' [51]. The mass shift can be split into normal mass shift (NMS) and a specific mass shift (SMS) components $MS = NMS + SMS$. The normal mass shift refers to the effect of the change in reduced mass $\frac{M_A m_e}{M_A + m_e}$ of the system, which alters the Bohr radius, a_0, in the classical orbit picture. The SMS appears in multi-electron systems due to the correlated motion of the electrons. This can have either a positive or negative contribution to the total mass shift. The SMS can be represented by the dot product expectation term $\left\langle \sum_{i>j} \vec{p}_i \cdot \vec{p}_j \right\rangle$ between two momentum vectors, summing over the electron momenta to calculate the binding energy of a specific electronic state. Together with the NMS this gives [51]

$$
\begin{aligned}
MS &= MS_{normal} + MS_{specific} , \\
&= \frac{M_{A'} - M_A}{(M_{A'} + m_e)(M_A + m_e)} \left(\frac{1}{2} \left\langle \sum_i p_i^2 \right\rangle + \left\langle \sum_{i>j} \vec{p}_i \cdot \vec{p}_j \right\rangle \right) , \\
&= \frac{M_{A'} - M_A}{(M_{A'} + m_e)(M_A + m_e)} (K_{NMS} + K_{SMS}) ,
\end{aligned}
\tag{2.34}
$$

Where the momenta can be written as K, a constant independent of the nuclear masses which contains the electronic structure information. The SMS term $\left\langle \sum_{i>j} \vec{p}_i \cdot \vec{p}_j \right\rangle$ presents a highly correlated many-body problem which is non trivial to calculate in multi-electron systems (see Sect. 2.3).

2.2.3.2 Field Shift

The shift in the electron energy levels depends on the electrostatic potential due to the density of nuclear charge, which is affected by the differences in the number of neutrons between isotopes as they rearrange the volume and shape of the nucleus. The electronic state is higher in energy for states which have higher probability density in the nuclear volume as they are less tightly bound in this region. The field shift effect is therefore largest for s electrons which have the highest spatial overlap with the nucleus.

The field shift contribution to the isotope shift increases with atomic number Z and nuclear mass A approximately as $\frac{Z^2}{A^{1/3}}$, due to the increase in electron density overlapping the nuclear volume, $V \propto A$, this dependence is shown in Fig. 2.5. The field-shift contribution therefore dominates over the mass shift for higher atomic numbers ($Z \gtrsim 40$).

There is an additional field-shift contribution from the shape of the nucleus, which is primarily due to the quadrupole deformation.

Fig. 2.5 Plot illustrating the relative contribution of field shift and mass shift with atomic number Z. The field-shift dependence approximately goes as $Z^2/^3\sqrt{A}$ while the mass shift goes as $1/A^2$. The offsets at $Z = 1$ from [50] were used. The field-shift contribution dominates over the mass-shift contribution for medium-mass nuclei ($Z \gtrsim 30$)

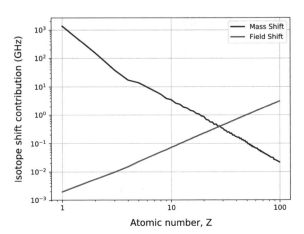

The mean-square charge radius $\langle r^2 \rangle$ is related to the nuclear deformation parameter, β_2, as in Eq. 2.10. In terms of the nuclear deformation parameter, β_2, and the volume, V, of the nucleus this becomes

$$V \simeq \frac{4}{3}\pi r_0^3 \left(1 + \frac{3}{4\pi}\beta_2^2\right) , \tag{2.35}$$

it is possible that even if the volume of the nucleus remained constant, the β_2 parameter would give rise to a change in $\langle r^2 \rangle$. The $\langle r^2 \rangle$ value therefore can have a contribution solely due to change in shape of the nucleus [51]. Although for most nuclei the two are correlated, the field shift in energy has contributions from both the volume shift and shape shift i.e. $FS = VS + SS$.

The field shift is calculated [52–54] by considering a change in binding energy for the overlapping electron probability density, which can be expressed as

$$\delta E_{FS} = F(Z)\lambda^{A,A'}$$
$$\simeq F(Z)\delta \langle r^2 \rangle^{A,A'} , \tag{2.36}$$

where $F(Z)$ is the field-shift factor which contains all terms needed to determine the change in isotope shift energy for a change in the nuclear parameter $\lambda^{A,A'}$, such as $\Delta|\psi_e(0)^2|$ the change in electron density near the nucleus between two levels in a transition [23]. The nuclear parameter $\lambda^{A,A'}$ is often approximated to the first-order radial moment $\delta \langle r^2 \rangle$. The higher-order radial moment expansion goes as

$$\lambda^{A,A'} = \delta \langle r^2 \rangle + \frac{C_2}{C_1}\delta \langle r^4 \rangle + \frac{C_3}{C_1}\delta \langle r^6 \rangle , \tag{2.37}$$

higher-order terms are typically important only for very heavy nuclei [50]. The Seltzer [55] coefficients (C_1, C_2, C_3) can be found in [56]. For indium these are $C_1 = 0.758 \times 10^2$, $C_2 = -0.462 \times 10^{-1}$ and $C_3 = 0.144 \times 10^{-2}$, meaning an expected contribution of less than 0.06% of the higher moments compared to $\delta \langle r^2 \rangle$. The higher order radial moments are thus neglected and included as an uncertainty in the extraction of charge radii in this work.

A field shift also occurs if an isomer has a different nuclear charge distribution than the ground state, an 'isomer shift'. This is particularly relevant in this work due to the abundance in isomers in the indium isotopes, which allowed the field shift of transitions to be evaluated with negligible contribution from the mass shift (Sect. 6.5).

2.2.3.3 King Plots

Using the relations above, the isotope shift in frequency is

$$\delta v_i^{A,A'} = M_i \overbrace{\frac{A'-A}{A'A}}^{\frac{1}{\mu_{A,A'}}} + F_i \lambda^{A,A'} . \tag{2.38}$$

By taking this relation for two transitions, i and j, and multiplying by the mass factor $\mu_{A,A'}$, so that $\mu_{A,A'}\lambda^{A,A'}$ can be cancelled, gives a linear equation for $\mu_{A,A'}\delta v_j^{A,A'}$ in terms of $\mu_{A,A'}\delta v_i^{A,A'}$ as

$$\mu_{A,A'}\delta v_j^{A,A'} = \overbrace{\frac{F_j}{F_i}}^{gradient} \mu_{A,A'}\delta v_i^{A,A'} + \overbrace{M_j - \frac{F_j}{F_i}M_i}^{intercept} . \tag{2.39}$$

By plotting pairs of isotope shift measurements in two separate transitions for each isotope (a 'King' plot [51]), a linear fit through the data with Eq. 2.39 allows atomic information to be extracted from the gradient and intercept values. The gradient, $\frac{F_j}{F_i}$, can be used to determine an unknown field shift constant F_j for a transition, if another F_i factor is already known.

2.3 The Quantum Many-Body Problem

While the solution to the full Schrödinger equation is not trivial in most cases, the quantum many-body states can readily be represented. The wavefunction for an A-body system $\Psi(x_1, x_2, x_3, \ldots, x_A)$ can be approximated as the product of single-particle states $\phi_1(x_1)$, $\phi_1(x_2)$ ...$\phi_1(x_A)$, if anti-symmetry is required this can be in the form of a Slater determinant [31] as

$$\Psi(x_1, x_2, \ldots, x_A) = \frac{1}{\sqrt{A!}} \det \begin{vmatrix} \phi_1(x_1) & \phi_1(x_2) & \ldots & \phi_1(x_A) \\ \phi_2(x_1) & \phi_2(x_2) & \ldots & \phi_2(x_A) \\ \vdots & & & \\ \phi_A(x_1) & \phi_A(x_2) & \ldots & \phi_A(x_A) \end{vmatrix} , \tag{2.40}$$

where $x_1, x_2 \ldots x_A$ represent the relevant quantum numbers. Slater determinants can be use to construct the many-body states for a Fermion system.

In full-configuration interaction (CI) calculations ('large shell model calculations' in the case of the nucleus) [57], A fermions are distributed over n single-particle states. By diagonalising the Hamiltonian matrix defined by the basis of all possible many-body states (Slater determinants), then the energy of the ground and all excited states can be solved, and similarly the expectation value of any other operator. Truncation of excitations, for example to single and double excitations (CISD) [58] can also be performed.

This approach is preferable if the dimensionality of the Hamiltonian matrix is small enough to solve computationally in a short amount of time. This is rarely the case for many quantum systems however and so alternative approaches are used. For the calculation of ground-state properties, powerful methods exist, such as the in-medium similarity re-normalisation group method (IM-SRG) [59–62], or the coupled-cluster (CC) method described below.

2.3.1 The Coupled-Cluster Method

The coupled-cluster method [63, 64] is based on the principle that by solving for a similarity transformation which makes a reference state without correlations $|\psi_0\rangle$, an eigenstate of the Hamlitonian $\bar{H}_N$

$$\bar{H}_N \equiv e^{-T} H_N e^T , \tag{2.41}$$

with T, the *cluster* operator used to introduce correlations into the state, $|\psi_c\rangle$.

$$T \equiv T_1 + T_2 + \cdots + T_A , \tag{2.42}$$

or using second-quantization formalism

$$T \equiv \sum_{ia} t_i^a a_a^\dagger a_i + \frac{1}{4} \sum_{ijab} t_{ij}^{ab} a_a^\dagger a_b^\dagger a_j a_i + \cdots + \frac{1}{(A!)^2} \sum_{i_1 \ldots i_A a_1 \ldots a_A} t_{i_1 \ldots i_A}^{a_1 \ldots a_A} a_{a_1}^\dagger \cdots a_{a_A}^\dagger a_{i_A} \cdots a_{i_1}. \tag{2.43}$$

Here $a_{a \ldots}^\dagger$ are the particle-creation operators and a_{i_A} are the hole-annihilation operators. This expansion can be approximated to $T \simeq T_1 + T_2$ (the first two terms), in the coupled-cluster singles and doubles approximation (CCSD). This significantly reduces the complexity of the problem, essentially from A-body to two-body. Typically 90% of the correlation energy is determined using up to doubles (CCSD), and 99% with triples [63] (CCSDT). The unknown cluster amplitudes t_i^a, t_{ij}^{ab} ... can then be found solving the coupled-cluster equations

$$\langle \Phi_0 | \overline{H_N} | \Phi_0 \rangle = E_c, \tag{2.44}$$

$$\langle \Phi_i^a | \overline{H_N} | \Phi_0 \rangle = 0, \tag{2.45}$$

$$\langle \Phi_{ij}^{ab} | \overline{H_N} | \Phi_0 \rangle = 0, \tag{2.46}$$

requiring that the reference state $|\psi_0\rangle$ to have no 1p-1h or 2p-2h excitations. Once the CC equations are solved for the cluster amplitudes, the ground-state energy of the system is then known by $E = E_c + E_{HF}$, where E_{HF} is the uncorrelated Hartree-Fock energy. Coupled-cluster theory is particularly powerful as $\overline{H}_N$ can be evaluated exactly [65], and using the CCSD approximation, the computational cost grows as $n_0^2 n_u^4$ instead of a factorial $n_0! n_u!$ as compared to the full configuration interaction (CI) calculation [63]. Extensions to isomeric states and even open quantum systems are also possible [66].

As an example, under a pairing or 'CCD' approximation, the author performed calculations for the correlation energy of a simple 8-level 4-particle system. The resulting correlation energy is shown in Fig. 2.6 for varying interaction strength g. Very good agreement was found with the full CI calculation, where the CCD calculation took only 10 s compared to around 5 min for the full CI. The coupled-cluster approach is widely applicable and is used in both atomic and nuclear structure calculations [66–69]. In order to find the solution for the coupled cluster equations e.g. Eqs. 2.45, 2.46 one must compute the matrix elements of the similarity-transformed Hamiltonian of Eq. 2.41, which can be performed using the Baker–Campbell–Hausdorf expansion:

$$\overline{H_N} = e^{-T} H_N e^T$$

$$= H_N + [H_N, T] + \frac{1}{2!} [[H_N, T], T] + \frac{1}{3!} [[[H_N, T], T], T] + \ldots . \tag{2.47}$$

One of the principle ingredients of the coupled-cluster approach which make it a highly accurate method and systematic approach to calculate correlations of a state is that the cluster operator (Eq. 2.43) is formed of sums of terms with particle creation and hole annihilation operators, but is absent of particle annihilation or hole creation operators. Therefore all of the terms which appear in T are commute and the only non-zero terms in Eq. 2.47 are those where T terms connect to $\overline{H_N}$ and not T with other T terms, therefore making the expansion finite. With this in mind, this also means the Baker–Campbell–Expansion can be formulated in terms of a diagrammatic expansion where topologically distinguishable diagrams for represent connecting terms in the expansion at a given order of approximation (see [63] for the rules to construct the diagrams). For example Fig. 2.7 shows the diagrams for

Fig. 2.6 Coupled-cluster calculation of correlation energy for a simple 4-particle, 8-orbits pairing model system. Even for a very strongly interacting system (large interaction strength g) the correlation energy was well reproduced with the coupled-cluster approximation compared to the full configuration interaction (CI). Calculations were performed in Python based on the CC formulas in [63]

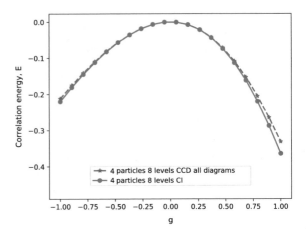

the example coupled-cluster doubles (CCD) approximation used for the calculations used to generate Fig. 2.6.[3]

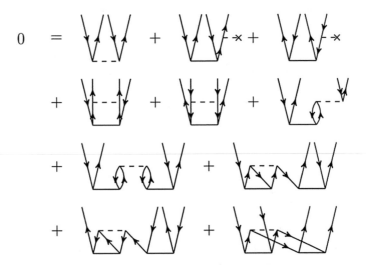

Fig. 2.7 The diagrammatic representation of the T_2 amplitude in the CCD approximation used for the calculation in Fig. 2.6. Figure taken from [70]. © 2020 American Physical Society[3]

[3]Reprinted from Baardsen, G. et al. Coupled-cluster studies of infinite nuclear matter. PRC 88, 054312 (2013). Copyright 2020 with permission from the American Physical Society.

2.3.2 Atomic Factor Calculations for Indium

Coupled-cluster CCSD(T) calculations were performed by [71] to determine the atomic parameters A_{hf}/g_I and B_{hf}/Q factors for the extraction of nuclear moments in this work. Where '(T)' denotes the inclusion of a perturbative correction for triple excitations [72, 73]. A comparison of the calculated factors to those determined experimentally is made in Sect. 5.5.

Along with correlation effects, relativistic effects play an important role in heavy-atom systems [68]. Hence a relativistic implementation was used for the indium atom calculations. A Dirac-Coulomb Hamiltonian [74] with a Dirac-Hartree-Fock (DHF) approach [75] was used to determine the mean-field wavefunction for the $[5s^2]$ closed configuration reference state (In$^+$) and the single-particle orbits. These were used in the relativistic coupled-cluster (RCC) calculations, which introduced the electron-correlation effects.

The hyperfine interaction operators [40] used to determine the hyperfine structure constants can be found in [72]. Higher-order corrections to the due to the Breit inter-action [76, 77] and quantum electrodynamic effects (QED) [78] were also included in the Hamiltonian. The A_{hf} values extracted experimentally in this work are compared to the calculated values in Sect. 5.5. A comparison was made between the values from the DHF and the CCSD approach, to demonstrate the necessity of electron-correlation effects, in addition to the triples, Breit and QED corrections.

CCSD(T) calculations of the FS and MS factors, F_i and $K_{MS,i} = K_{NMS,i} + K_{SMS,i}$ respectively, were also performed by [79], allowing accurate extraction of $\delta \langle r^2 \rangle^{A,A'}$ from the isotope shift measurements, for the transitions, i, used in this work. For the calculation of these factors an analytic-response (AR) RCC approach was used [80, 81]. Although the AR approach had already existed [67], the calculations were applied to calculate the isotope shift factors in an atomic system for the first time. The AR approach was particularly appropriate in the calculation of the SMS factor, $K_{SMS,i}$, as it avoided the appearance of non-terminating series which appear when evaluating the two-body SMS operator using the expectation-value-evaluation (EVE) approach [52, 54, 82–84], which was used for the hyperfine interaction constants [79]. Calculations using a finite field (FF) approach were also performed by [71] for comparison. Compared to AR approach, the finite field approach has the drawback of a dependence on the choice of a perturbative parameter [52, 67].

This AR approach allowed the MS factors of atomic indium to be calculated for the first time and offers an accurate approach to calculate the SMS factor for other odd-proton systems, which have the common problem described in Sect. 6.5 for isotone independent extraction of charge radii from isotope shift measurements. Comparison of the calculated FS and MS factors to experimentally extracted values is given in Sect. 6.5.1.

2.4   Atomic Populations Following Charge Exchange

Many collinear laser spectroscopy (CLS) measurements are performed on atomic rather than ionic beams, in part due to the increased availability of spectroscopic data but also, in the case of CRIS, due to the lower ionization potential of atomic systems. This reduces the technical complexity of step-wise laser ionization. Producing a neutral atom beam is typically performed through electron-capture reactions with an alkali vapour. During these reactions many atomic states may be populated. The prediction of this distribution for ion beam energies in the 1–100 keV regime is discussed in this section.

Populating many atomic states during charge exchange significantly reduces the total efficiency of CLS measurements performed from only one of the atomic states of the neutralised beam. For the work on the indium isotope chain, the population distribution was carefully measured during the laser scheme development phase of this project. The results from this work are discussed in Sect. 5.9. The design of the charge-exchange cell used at CRIS is shown in Sect. 3.2.1.

2.4.1   Neutralisation Reactions

An ion beam composed of ions, A^+, can be neutralised by capturing an electron from a free atom B. The atom B is typically an alkali metal (e.g. potassium or sodium), as they have a low-IP and low boiling point, increasing the ease of vapour formation. In neutralisation reactions between ions and atoms of the same element $B^+ + B \rightarrow B + B^+$, i.e. symmetric charge exchange, the probability of electron capture is high and decreases with increasing beam energy, T_B, as

$$\sigma_{CE}^{1/2} = a - b \ln T_B \, , \tag{2.48}$$

where a and b are empirical constants, their values can be extracted from cross sections for charge exchange measured for a given element [85–88]. In the case of a symmetric reaction and low beam energies ($\lesssim 1$ keV), the population of the atomic ground state following neutralisation of the ion will be largest. For many CLS experiments, the atomic number of the ion to be neutralised and the atomic number of the free atom from which it will capture an electron will be different, these asymmetric charge-exchange reactions have the form

$$A^+(0) + B(0) \rightarrow A(k) + B^+(0) + \Delta E \, . \tag{2.49}$$

Where $B(0)$ is the free atom in its ground state, and $A^+(0)$ is an ion from the beam which captures an electron from $B(0)$, and then exits in an excited state, k. The excited state, k, depends on the difference in ionization potential of the two elements in the reaction, $\Delta E = I_A - I_B$, and can therefore be much above the ground state of the product atom. With the typical beam energies used for collinear laser spectroscopy (>1 keV), this energy deficit, ΔE, is not a significant barrier to the reaction, and therefore many high-lying states can be populated in these experiments.

The maximum cross section σ_{CE} for an asymmetric charge-exchange reaction with energy deficit ΔE can approximately be found at a velocity

$$v \simeq \frac{a\Delta E}{h} . \tag{2.50}$$

This follows from the Massey hypothesis [89], which assumes an 'adiabatic' velocity region, where the reduction in internucleon distance from the relative velocity is much slower than the electronic motion, allowing electronic transitions to be avoided. Following the maximum σ_{CE} at velocity, v, the σ_{CE} will decrease with increasing beam energy, T_B. The adiabatic parameter value of $a = 7 \times 10^{-8}$ cm has been found to have agreement between several reaction experiments [88, 90–92]. For collinear laser spectroscopy experiments, the cross sections for populating excited states, k, is of interest, as they form the initial atomic population distribution of the atom which measurements are performed from. An atomic-energy-level picture of the charge exchange process can be seen in Fig. 2.8, for indium ions neutralised by sodium, the final population distribution after spontaneous de-excitation over 120 cm of atom flight (see Sect. 2.4.3) is also shown, at $T_B = 20$ keV and $T_B = 40$ keV.

As suggested by Eq. 2.50, the velocity can be tuned for state-selective neutralisation, allowing the neutralisation process to be used for ultrasensitive radioisotope detection, the relative cross sections can be further manipulated for state-selective neutralisation by optically pumping of the reactant atom or ion [93, 94].

The total cross section for neutralisation is difficult to calculate as it can proceed through multiple reaction channels, and is heavily dependent on the reactant atom and ion elements, $A^+(i), B(j)$. However an approximate method was developed by Rapp and Francis [95] (R&F), using a semi-classical impact parameter approach, which allows calculation of the cross section for the population of excited states k of product atoms in asymmetric charge exchange reactions. This method was used to calculate the neutralisation cross sections in the following section.

2.4.2 Neutralisation Cross Section Calculation

In the impact-parameter approach, a three-body approximation is used for the electron in the field of the projectile and target ions, the dynamics of this three-body system are found determined as a function of time t by the incident velocity v, and impact parameter b, values. The probability, $P(b, v)$, of the electron capture is then also a function of b and v values, which when integrated over for the impact parameters gives the cross section as a function of the incident velocity, v and the relevant atomic parameters as

$$\sigma(b, v) = 2\pi \int_0^\infty P_\omega(b, v) b \, db \, . \tag{2.51}$$

The probability for asymmetric charge exchange, $P_\omega(b, v)$, is given by the R&F method as

$$P_\omega(b, v) = f \, P_0(b, v) \, \mathrm{sech} \left(\frac{\omega}{v} \sqrt{\frac{a_0 \pi b}{2\gamma}} \right)^2 , \tag{2.52}$$

where $P_0(b, v)$ is the symmetric charge-exchange probability, a_0 is the Bohr radius, $\gamma = \sqrt{\frac{\bar{I}}{13.6 \, \text{eV}}}$ and $\bar{I} = (I_A - I_B)/2$. To take into account ratio of multiplicity between the products orbital angular momenta and spin and that of the reactants, a statistical factor f is included, accounting for the fraction of collisions which are able to form the product spin states. The factor consists of degeneracies of the spin f_S and orbital angular momenta f_L of the reactant and product states $f = f_S f_L$. This is given by the ratio of the multiplicity $(2S + 1)$ of the spin state formed with correct spin of the products to the multiplicity of all spin states that can be formed from the reactants spin. For f_S this is given by

$$f_s = \frac{\text{products multiplicities}}{\text{reactants multiplicities}} = \frac{(2S_{A(k)} + 1)(2S_{B^+(0)} + 1)}{(2S_{B(0)} + 1)(2S_{A^+(0)} + 1)} , \tag{2.53}$$

for example reactants with $S = \frac{3}{2}$ and $S = \frac{1}{2}$ would have a weight factor of $\frac{2 \times 1 + 1}{(2 \times 3/2 + 1)(2 \times 1/2 + 1)} = \frac{3}{8}$ for forming a state with $S = 1$. The same logic applies to orbital angular momenta L degeneracies when the reactants have higher angular momenta than the products. Otherwise the angular momentum can be gained from the collision, $f_L = 1$ is used in these simulations when this is the case (if $f_L > 1$).

The contribution from the probability for symmetric charge exchange is defined [95] as

$$P_0(b, v) = \sin \left(\sqrt{\frac{2\pi}{\gamma a_0} \frac{2\bar{I}}{\hbar v}} b^{\frac{3}{2}} \left(1 + \frac{a_0}{\gamma b} \right) \exp(\frac{-\gamma b}{a_0}) \right)^2 , \tag{2.54}$$

where the average ionization potential of the atoms in the asymmetric reaction, $\bar{I} = \frac{I_A + I_B}{2}$, is used. From the work by [96], a correction to the original formula by a factor of 2 is included within the sin function. A linear trajectory before and after electron pick up is assumed in the impact parameter approach [97, 98]. While this assumption holds in the majority of cases, there is in reality an angular distribution of the product neutrals. In an extreme case of $O^+ + N_2$ [99] at 0.5−5 keV, a small fraction of the neutral beam was found at angles of up to 3°, which was attributed to closer collision distances due to a large energy deficit.

Equation 2.54 is derived under the assumption of an 'intermediate' velocity region, defined [100] as

$$m_B \text{ meV} < T_B < 10 \, m_B \text{ keV} , \qquad (2.55)$$

where the incident ion mass, m_B, is given in atomic mass units. For incident ion beam energies typical of CLS experiments (1–100 keV) all incident elements fall under this range, apart from $1 \leq Z \leq 3$ which are on the boundary.

The neutralisation cross sections for the neutralisation of indium ions to all experimentally known product electronic states in atomic indium was calculated, using the spectroscopic level information from the NIST atomic database [101], and the above R&F approach. The same calculations were also performed for elements $1 \leq Z \leq 89$, the results are discussed in Sect. 5.9.

In the original formulation by R&F [95], the approximation of $P_0(b_1, v) = \frac{1}{2}$ is used up to a 'cut-off' impact parameter, b_1. In this work numerical integration of Eq. 2.54 was used up to a convergence level instead, using a Levin-type [102] integration rule (for oscillatory integrals) implemented in Mathematica® [103]. This avoided solving for b_1 for each electronic state considered.

Many contributions included in more recent and complete approaches [104–107] are neglected in the R&F [95] method. For instance the impact parameter approach considers only the non-radiative kinematic mechanism [97, 108] for electron capture. However radiative channels have also been identified to contribute to first order in electron-transfer mechanisms, where photon emission [109] or electron emission [110] also accompanies the electron capture.

Experimental data has been compiled by [111] for many symmetric charge-exchange reactions, in which it was found that the R&F approach has poorer agreement at T_B <5 keV/u for atoms with ionization potentials much smaller than hydrogen. The correction needed for the calculations to reproduce the cross-section data was found [112, 113] and a corrective formula is reported in [114]. Although this correction allows for in increased accuracy of the calculations, CLS experiments are rarely performed at T_B <5 keV due to the undesirable increase in the Doppler broadening of the spectra. To increase the accuracy of the method for symmetric charge exchange by double-electron capture [115, 116], an improvement to the single electron potential used by R&F has been reported by [115] which improved the agreement with experimentally measured cross sections for reactions of this kind. Where more accurate calculation of cross sections for double-electron capture reactions are needed, approaches have recently been developed which correctly include four-body interactions [117–119].

Although the absolute cross sections calculated from the R&F method are known to disagree with experiment values in some cases [112, 116, 120, 121], the calculations well reproduce the dependence of cross section on ion velocity for asymmetric charge exchange. For calculation of relative populations, where the cross section for neutralisation to thousands of atomic states has to be calculated for some elements, the R&F method proves to be a computationally inexpensive approach and has been shown predict final relative atomic populations which agree with experiment [122,

123]. The largest source of error for the relative population calculations in most atomic systems can be attributed to incomplete spectroscopic data for the element. This is discussed in Sect. 5.9.[4]

2.4.3 Relative Atomic Populations

Following calculation of the cross sections for charge exchange, σ_{CE}, for each atomic state, the de-excitation of the population to lower atomic levels takes place over a distance l_{flight} at velocity v. The de-excitation was simulated using the spontaneous decay coefficients, A_{ki}, taken from the NIST atomic database [101]. Missing A_{ki} values represent a large source of uncertainty in the calculation, where transitions with unknown A_{ki} values were not used in the simulation. Although this source of error is difficult to evaluate, it will generally result in overestimation of high-lying state populations and underestimation of low-lying populations with respect to the difference in IPs of the elements involved in the reaction. Branching ratios, $\beta_{ki'}$, were calculated from the coefficients as $\beta_{ki'} = \frac{A_{ki'}}{\sum_i A_{ki}}$, to lower states, i.

The population decay from upper states, k, were then evolved using

$$F_k^{decay} = \exp\left(-\frac{l_{flight}}{v} \sum_i A_{ki}\right). \tag{2.56}$$

The decay paths through intermediate states were tracked at each iteration of the simulation.

Using this approach, the relative populations of indium neutralised by the reaction

$$\text{In}^+(0) + \text{Na}(0) \rightarrow \text{In}(k) + \text{Na}^+(0) + \Delta E \tag{2.57}$$

with sodium and with potassium

$$\text{In}^+(0) + \text{K}(0) \rightarrow \text{In}(k) + \text{K}^+(0) + \Delta E \tag{2.58}$$

were simulated. The results are displayed in Fig. 2.8A. An atom flight distance of $l_{fl} = 120$ cm was used for the final populations. A comparison is shown between ion beam energies of $T_B = 40$ keV and $T_B = 20$ keV.

From these simulations it was clear that the use of sodium as the alkali metal vapour for charge exchange would give the largest population of the metastable $^2P_{3/2}$ state at 2212.6 cm^{-1}. The $^2P_{3/2}$ state being the lower state of the 246.8-nm (5p $^2P_{3/2} \rightarrow$ 9s $^2S_{1/2}$) transition used to give sensitivity to the quadrupole moment Q_s, the ground state being $^2P_{1/2}$. Figure 2.9 shows the case of using indium itself

---

[4]Reprinted from Vernon, A. R. et al. Simulation of the relative atomic populations of elements $1 \leq Z \leq 89$ following charge exchange tested with collinear resonance ionization spectroscopy of indium. Spectrochim. Acta Part B At. Spectrosc. 153, 61–83, Copyright (2019) with permission from Elsevier.

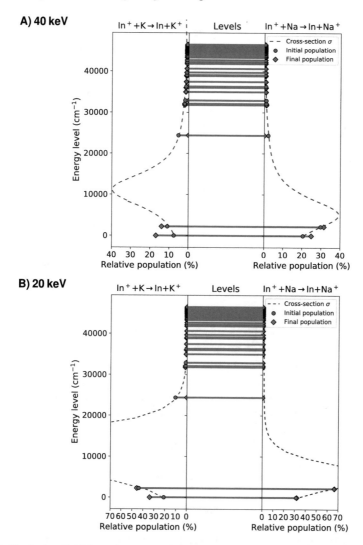

Fig. 2.8 The final and initial atomic state populations of atomic indium, compared for neutralisation by sodium or potassium, as well as for (**A**) T_B =40 keV and (**B**) T_B = 20 keV. A flight length of l_{fl} = 120 cm for the atom was used. The $^2P_{3/2}$ state has the highest relative population, with a cross section of $\sigma(^2P_{3/2}) = 3.93 \times 10^{-16}$ cm^2 using potassium or $\sigma(^2P_{3/2}) = 4.17 \times 10^{-15}$ cm^2 using sodium at $T_B = 20$ keV.[4]

to neutralise the In$^+$ beam. The conventional approach in many CLS experiments is simply to consider the energy balance for the reaction, in which case symmetric neutralisation reactions would be the ideal case. As Fig. 2.9 indicates however, this does not take account of any other states in the atom, other than the state of interest. In this case using indium to neutralise the In$^+$ ions would result in a lower relative population of the $^2P_{3/2}$ state of interest, in addition to the impracticality of creating an indium vapour.

Fig. 2.9 The final and initial atomic state populations of atomic indium, compared for neutralisation by sodium or indium itself at $T_B = 40$ keV. For comparison the cross section for populating the $^2P_{3/2}$ state using sodium is $\sigma(^2P_{3/2}) = 5.05 \times 10^{-15}$ cm^2 and $\sigma(^2P_{3/2}) = 4.17 \times 10^{-15}$ cm^2 using indium

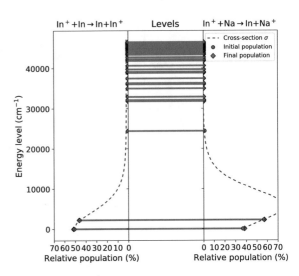

These relative populations were later measured at $T_B = 20$ keV with naturally occuring indium isotopes with an ablation ion source, which is the topic of Sect. 5.9. This was used as a evaluation of the populations for the experiment on radioactive indium isotopes which was later performed at $T_B = 40$ keV, shown in Fig. 2.8A.

The relationship between the relative population of a state and its cross section as a function of T_B depends highly on the atomic levels of the elements of interest for the reaction. Take the 'off-resonance' asymmetric reaction

$$\text{In}^+(0) + \text{Na}(0) \rightarrow \text{In}(2212.6 \text{ cm}^{-1}) + \text{Na}^+(0) , \qquad (2.59)$$

and compare it to the 'resonant' symmetric reaction

$$\text{Na}^+(0) + \text{Na}(0) \rightarrow \text{Na}(0) + \text{Na}^+(0) . \qquad (2.60)$$

For beam energies, $T_B = 1$–100 keV, this is shown in Fig. 2.10. For the symmetric reaction, both the cross section for neutralising into and relative population of the ground state of sodium simply decrease with beam energies. The cross section follows the rule shown in Eq. 2.50 and is well established experimentally [95, 111]. As T_B increases, the contribution from other states to the final population increases, and therefore the difference between final and initial populations also increases.

This is in contrast the the relationship between the relative population and cross section for neutralising into the 5p $^2P_{3/2}$ state of indium at (2212.6 cm^{-1}), due to the difference in ionization potential with sodium. The broadening of the 'resonance' with increasing T_B, shown in Fig. 2.8, is key to populating the 5p $^2P_{3/2}$ state.

Hence when choosing a reactant atom for neutralisation and an optimum beam energy to maximise the population of a given atomic state, one has to take these factors

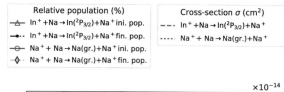

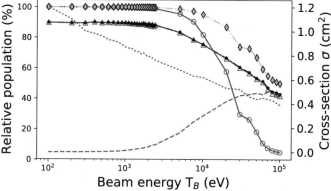

Fig. 2.10 Initial and final electronic state populations and cross sections for the reactions $In^+(0) + Na(0) \rightarrow In(2212.6\ cm^{-1}) + Na^+(0)$ and $Na^+(0) + Na(0) \rightarrow Na(0) + Na^+(0)$. For beam energies in the range $T_B = 0$–$100\ keV$ and with de-excitation following an atom flight length of $l_{fl} = 120\ cm$. [4]

into account. Additionally, the total neutralisation cross-sections (shown in Table B.3) may increase with T_B due to the broadening, in practice this can be compensated for by increased vapour pressure however and is therefore not an intrinsic limitation to the population of a state.

References

1. Sakurai JJJJ et al (2011) Modern quantum mechanics. Addison-Wesley, Boston
2. Otsuka T, et al (2010) Phys Rev Lett 105(3):032501. ISSN 0031-9007. https://doi.org/10.1103/PhysRevLett.105.032501
3. Roth R, et al (2011) Phys Rev Lett 107(7):072501. ISSN 0031-9007. https://doi.org/10.1103/PhysRevLett.107.072501
4. Jurgenson ED, et al (2011) Phys Rev C 83(3):034301. ISSN 0556-2813. https://doi.org/10.1103/PhysRevC.83.034301
5. Krane KS (2008) Introductory Nuclear Physics. Wiley, New Delhi, India
6. Woods RD, et al (1954) Phys Rev 95(2):577. ISSN 0031-899X. https://doi.org/10.1103/PhysRev.95.577
7. Mayer MG (1950) Phys Rev 78(1):16. ISSN 0031-899X. https://link.aps.org/doi/10.1103/PhysRev.78.16
8. Suhonen J (2007) Nuclear two-body interaction and configuration mixing. Springer, Berlin, Heidelberg. https://doi.org/10.1007/978-3-540-48861-3_8
9. Kim YH, et al (2018) Phys Rev C 97(4):041302. ISSN 2469-9985. https://doi.org/10.1103/PhysRevC.97.041302
10. Casten R (2000) Nuclear structure from a simple perspective. Oxford University Press, Oxford

11. Bacca S, et al (2014) J Phys G: Nucl Part Phys 41(12):123002. ISSN 0954-3899. http://stacks. iop.org/0954-3899/41/i=12/a=123002?key=crossref.b94b55546a16f222baab82846d1a357d
12. Beloy K, et al (2008) Phys Rev A 77(5):052503. ISSN 1050-2947. https://doi.org/10.1103/ PhysRevA.77.052503
13. Gerginov V, et al (2003) Phys Rev Lett 91(7):072501. ISSN 0031-9007. https://doi.org/10. 1103/PhysRevLett.91.072501
14. Eck TG, et al (1957) Phys Rev 106(5):958. ISSN 0031-899X. https://doi.org/10.1103/ PhysRev.106.958
15. Sheng D, et al (2010) J Phys B: Atom Mol Opt Phys 43(7):074004. ISSN 0953-4075. http://stacks.iop.org/0953-4075/43/i=7/a=074004?key=crossref. a441ec0e93686dc7abf232f58fc94489
16. Hammen M, et al (2018) Phys Rev Lett 121(10):102501. ISSN 0031-9007. https://doi.org/ 10.1103/PhysRevLett.121.102501
17. Reinhard P-G, et al (2017) Phys Rev C 95(6):064328. ISSN 2469-9985. https://doi.org/10. 1103/PhysRevC.95.064328
18. Garcia Ruiz RF, et al (2016) Nat Phys 12(6):594. ISSN 1745-2473. http://www.nature.com/ articles/nphys3645, https://doi.org/10.1038/nphys3645
19. Weizsacker CFv (1935) Zeitschrift fur Physik, 96(7–8):431. ISSN 1434-6001. https://doi.org/ 10.1007/BF01337700
20. Poenaru DN, et al (1996) Handbook of nuclear properties. Clarendon Press, Wotton-under-Edge. https://books.google.ch/books/about/Handbook_of_Nuclear_Properties.html? id=nzN1HBXOsVQC&source=kp_cover&redir_esc=y
21. Rohozinski SG, et al (1980) J Phys G: Nucl Phys 6(8):969. ISSN 0305-4616. http://stacks. iop.org/0305-4616/6/i=8/a=008?key=crossref.1b9ce50c9e35a00a8d215d6accc348a1
22. Myers WD, et al (1983) Nucl Phys A 410(1):61. ISSN 0375-9474. https://www.sciencedirect. com/science/article/pii/0375947483904013
23. Campbell P, et al (2016) Prog Part Nucl Phys 86:127. ISSN 01466410. https://doi.org/10. 1016/j.ppnp.2015.09.003, https://linkinghub.elsevier.com/retrieve/pii/S0146641015000915
24. Neyens G (2003) Rep Prog Phys 66(4):633. ISSN 0034-4885. http://stacks.iop.org/0034-4885/66/i=4/a=205?key=crossref.a8f63ecd4735d553dad9fc7c51e39bae
25. NIST (2018) CODATA value: nuclear magneton in MHz/T. https://physics.nist.gov/cgi-bin/ cuu/Value?munshhz%7Csearch_for=nuclear+magneton
26. Bohr A et al (1998) Nuclear structure. World Scientific, Singapore
27. Heyde K (1988) Hyperfine Interact 43(1–4):15. ISSN 0304-3843. https://doi.org/10.1007/ BF02398284
28. Heyde KLG (1990) The nuclear shell model. Springer series in nuclear and particle physics. Springer, Berlin, Heidelberg. https://doi.org/10.1007/978-3-642-97203-4, https://books.google.ch/books?id=aBz4CAAAQBAJ&dq=heyde+nuclear+shell+model& source=gbs_navlinks_s
29. Stone N (2016) Atom Data Nucl Data Tables 111–112:1. ISSN 0092-640X. https://www. sciencedirect.com/science/article/pii/S0092640X16000024
30. Castel B, et al (1990) Modern theories of nuclear moments. Clarendon Press, Oxford. https://books.google.fr/books/about/Modern_Theories_of_Nuclear_Moments.html? id=os8XAAAAIAAJ&redir_esc=y
31. Fényes T (2002) Structure of atomic nuclei. Akadémiai Kiadó
32. Sagawa H, et al (1988) Phys Lett B 202(1):15. ISSN 0370-2693. https://www.sciencedirect. com/science/article/pii/0370269388908453
33. Rowe DJ (2010) The collective rotational model. World Scientific, Singapore. https://doi.org/ 10.1142/9789812790668_0006
34. Heyde K (2004) Basic ideas and concepts in nuclear physics: an introductory approach, 3rd edn. CRC Press, Boca Raton
35. Berdichevsky D, et al (1985) Z Phys A-Atoms Nuclei 322(1):141. ISSN 0340-2193. https:// link.springer.com/content/pdf/10.1007%2FBF01412027.pdf

36. Wilkins SG, et al (2017) Phys Rev C 96(3):034317. ISSN 2469-9985. https://doi.org/10.1103/PhysRevC.96.034317
37. Stone N (2005) Atom Data Nucl Data Tables 90(1):75. ISSN 0092640X. http://linkinghub.elsevier.com/retrieve/pii/S0092640X05000239
38. Timar J, et al (2014) Nucl Data Sheets 121:143. ISSN 0090-3752. https://www.sciencedirect.com/science/article/pii/S0090375214006565
39. Hashizume A (2011) Nucl Data Sheets 112(7):1647. ISSN 0090-3752. https://www.sciencedirect.com/science/article/pii/S009037521100055X
40. Schwartz C (1955) Phys Rev 97(2):380. ISSN 0031-899X. https://doi.org/10.1103/PhysRev.97.380
41. Flynn CP, et al (1960) Proc Phys Soc 76(2):301. ISSN 0370-1328. http://stacks.iop.org/0370-1328/76/i=2/a=415?key=crossref.f3515cb119eec3150e31da423e1fdb38
42. Shore BW et al (1967) Principles of atomic spectra. Wiley, Hoboken
43. Shore BW et al (1967) Principles of atomic spectra. Wiley, Hoboken
44. Rosenthal JE, et al (1932) Phys Rev 41(4):459. ISSN 0031-899X. https://doi.org/10.1103/PhysRev.41.459
45. Rosenthal JE, et al (1932) Phys Rev 41(4):459. ISSN 0031-899X. https://doi.org/10.1103/PhysRev.41.459
46. Rosenberg HJ, et al (1972) Phys Rev A 5(5):1992. ISSN 0556-2791. https://doi.org/10.1103/PhysRevA.5.1992
47. Frommgen N, et al (2015) Eur Phys J D 69:164. http://www.eli-np.ro/scientific-papers/Collinearlaserspectroscopyofatomiccadmium.pdf
48. Eck TG, et al (1957) Phys Rev 106(5):954. ISSN 0031-899X. https://doi.org/10.1103/PhysRev.106.954
49. Persson JR (2013) Atom Data Nucl Data Tables 99(1):62. ISSN 0092640X. https://www.sciencedirect.com/science/article/pii/S0092640X1200085X
50. Nörtershäuser W, et al (2014) The Euroschool on exotic beams, Vol. IV. No. 529 in lecture notes in physics. Springer, Berlin, Heidelberg. https://doi.org/10.1007/978-3-642-45141-6
51. King WH (2013) Isotope shifts in atomic spectra, Springer Science & Business Media, vol. 11. https://books.google.com/books?id=eEgGCAAAQBAJ&pgis=1
52. Sahoo BK (2010) J Phys B: Atomic Mol Opt Phys 43(23):231001. http://stacks.iop.org/0953-4075/43/i=23/a=231001?key=crossref.b347b259dc6e594fa0b235601393df70
53. Safronova MS, et al (2001) Phys Rev A 64(5):052501. ISSN 1050-2947. https://doi.org/10.1103/PhysRevA.64.052501
54. Berengut JC, et al (2003) Phys Rev A 68(2):022502. ISSN 1050-2947. https://doi.org/10.1103/PhysRevA.68.022502
55. Seltzer EC (1969) Phys Rev 188(4):1916. https://doi.org/10.1103/PhysRev.188.1916
56. Boehm F, et al (1974) Atomic Data Nucl Data Tables 14(5–6):605. ISSN 0092-640X. https://www.sciencedirect.com/science/article/pii/S0092640X74800057
57. Brown B (2001) Prog Part Nucl Phys 47(2):517. ISSN 0146-6410. https://www.sciencedirect.com/science/article/pii/S0146641001001594?via%3Dihub
58. Grev RS, et al (1992) J Chem Phys 96(9):6850. ISSN 0021-9606. https://doi.org/10.1063/1.462574
59. Parzuchowski NM, et al (2017) Phys Rev C 96(3):034324. ISSN 24699993
60. Hergert H (2016) arXiv:1607.06882
61. Wegner F (1994) Annalen der Physik 506(2):77. ISSN 00033804. https://doi.org/10.1002/andp.19945060203
62. Głazek SD, et al (1993) Phys Rev D 48(12):5863. ISSN 0556-2821. https://doi.org/10.1103/PhysRevD.48.5863
63. Hjorth-Jensen M, et al (2017) Computational nuclear physics and post hartree-fock methods. Springer, Berlin, Heidelberg. https://doi.org/10.1007/978-3-319-53336-0_8
64. Wilson KG (1990) Nucl Phys B - Proc Suppl 17:82. ISSN 0920-5632. https://www.sciencedirect.com/science/article/pii/092056329090223H?via%3Dihub

65. Achilles R, et al (2012) Arch History Exact Sci 66(3):295. ISSN 0003-9519. https://doi.org/10.1007/s00407-012-0095-8

66. Hagen G, et al (2007) Phys Lett B 656(4–5):169. ISSN 0370-2693. https://www.sciencedirect.com/science/article/pii/S0370269307010593

67. Bishop RF (1991) Theor Chim Acta 80(2–3):95. ISSN 0040-5744. https://doi.org/10.1007/BF01119617

68. Kaldor U (1995) Relativistic coupled cluster calculations. Springer US, Boston, MA. https://doi.org/10.1007/978-1-4615-1937-9_13

69. Garcia Ruiz RF, et al (2015) Phys Rev C 91:041304. ISSN 0556-2813

70. Baardsen G, et al (2013) Phys Rev C - Nucl Phys 88(5):054312. ISSN 1089490X

71. Sahoo BK (2018) Private communication. Breit and QED effects, CCSD(T) calculations for indium inclusing higher order relativistic effects

72. Garcia Ruiz RF, et al (2018) CERN-INTC-2018- P546. https://cds.cern.ch/record/2299760/files/INTC-P-546.pdf

73. Riplinger C, et al (2013) J Chem Phys 139(13):134101. ISSN 0021-9606. https://doi.org/10.1063/1.4821834

74. Ilyabaev EE, et al (1994) Chem Phys Lett 222(1–2):82. ISSN 0009-2614. https://www.sciencedirect.com/science/article/pii/0009261494003173

75. Visscher L (2002) Theoret Comput Chem 11:291. ISSN 1380-7323. https://www.sciencedirect.com/science/article/abs/pii/S1380732302800322

76. Mann JB, et al (1971) Phys Rev A 4(1):41. ISSN 0556-2791. https://doi.org/10.1103/PhysRevA.4.41

77. Breit G (1929) Phys Rev 34(4):553. ISSN 0031-899X. https://doi.org/10.1103/PhysRev.34.553

78. Sahoo BK (2016) Phys Rev A 93(2):022503. ISSN 2469-9926. https://doi.org/10.1103/PhysRevA.93.022503

79. Sahoo B, et al (2019) Phys Rev Lett

80. Szalay PG (1995) Int J Quantum Chem 55(2):151. ISSN 0020-7608. https://doi.org/10.1002/qua.560550210

81. Monkhorst HJ (1977) Int J Quantum Chem 12(S11):421. ISSN 00207608. https://doi.org/10.1002/qua.560120850

82. Cubiss JG, et al (2018) Phys Rev C 97:21. https://doi.org/10.1103/PhysRevC.97.054327

83. Wansbeek LW, et al (2012) Phys Rev C 86(1):015503. ISSN 0556-2813. https://doi.org/10.1103/PhysRevC.86.015503

84. Gaidamauskas E, et al (2011) J Phys B: Atomic Mol Opt Phys 44(17):175003. ISSN 0953-4075. http://stacks.iop.org/0953-4075/44/i=17/a=175003?key=crossref.bd8f33975d7c2caf5cd88bc03ae7edc4

85. Dalgarno A, et al (1958) Philos Trans R Soc A: Math Phys Eng Sci 250(982):426. ISSN 1364-503X. https://doi.org/10.1098/rsta.1958.0003

86. Demkov Y (1952) Quantum-mechanical calculation of the probability of charge-exchange in collisions. Leningrad State University

87. Dalgarno A, et al (1957) Philos Trans R Soc A: Math Phys Eng Sci 240(1221):284. ISSN 1364-5021. https://doi.org/10.1098/rspa.1957.0084

88. Lindsay BG, et al (2005) J Geophys Res: Space Phys 110(A12):1. ISSN 21699402

89. Massey HSW (1949) Rep Prog Phys 12(1):311. ISSN 00344885. http://stacks.iop.org/0034-4885/12/i=1/a=311?key=crossref.69f2667cda91e143096f1e5fb17fec47

90. Sycheva AA, et al (2016) J Phys Chem A 120(27):4655. ISSN 1089-5639. https://doi.org/10.1021/acs.jpca.5b09151

91. Hasted JB (1972) Phys Atomic Coll Am Elsevier, 2nd edn. https://bibdata.princeton.edu/bibliographic/1917772

92. Stedeford JBH, et al (1955) Proc R Soc A: Math Phys Eng Sci 227(1171):466. ISSN 1364-5021. https://doi.org/10.1098/rspa.1955.0024

93. Vermeeren L, et al (1992) Phys Rev Lett 68(11):1679. ISSN 0031-9007. https://doi.org/10.1103/PhysRevLett.68.1679

94. Garcia Ruiz R, et al (2017) J Phys G: Nucl Part Phys
95. Rapp D, et al (1962) J Chem Phys 37(11):2631. ISSN 00219606. https://doi.org/10.1063/1.1733066
96. Dewangan DP (1973) J Phys B: Atomic Mol Phys 6(2):L20. ISSN 0022-3700. https://doi.org/10.1088/0022-3700/6/2/004
97. Brinkman H et al (1930) Proc. Acad. Sci. Amsterdam 8:973
98. Gurnee EF, et al (1957) J Chem Phys 26(5):1237. ISSN 0021-9606. https://doi.org/10.1063/1.1743499
99. Lindsay BG, et al (1998) Phys Rev A 57(1):331. ISSN 1050-2947. https://doi.org/10.1103/PhysRevA.57.331
100. Bates DR, et al (1953) Proc R Soc A: Math Phys Eng Sci 216(1127):437. ISSN 1364-5021. https://doi.org/10.1098/rspa.1953.0033
101. Kramida NATA, Ralchenko Y, Reader J (2014) NIST Atomic Spectra Database. http://www.nist.gov/pml/data/asd.cfm
102. Levin D (1996) J Comput Appl Math 67(1):95. ISSN 03770427
103. Hascelik AI, et al (2014) Appl Math Sci 8(138):6889. www.m-hikari.com, https://doi.org/10.12988/ams.2014.49720
104. Tolstikhina IY, et al (2012) J Phys B: Atomic Mol Opt Phys 45(14):145201. ISSN 0953-4075. http://stacks.iop.org/0953-4075/45/i=14/a=145201?key=crossref.228d8432aab161dea37dbf937b6bdd3a
105. Dewangan DP, et al (1994) Phys Rep 247(2–4):59. ISSN 03701573
106. Bransden BH et al (1992) Charge exchange and the theory of ion-atom collisions. Clarendon Press, Oxford
107. Dewangan D, et al (1987) Nucl Instrum Methods Phys Res Sect B: Beam Interact Mater Atoms 23(1–2):160. ISSN 0168-583X. https://www.sciencedirect.com/science/article/pii/0168583X8790437X
108. Oppenheimer JR (1928) Phys Rev 31(3):349. ISSN 0031-899X. https://doi.org/10.1103/PhysRev.31.349
109. Briggs JS, et al (1974) Phys Rev Lett 33(19):1123. ISSN 0031-9007. https://doi.org/10.1103/PhysRevLett.33.1123
110. Najjari B, et al (2008) Phys Rev Lett 101(22):1. ISSN 00319007
111. Sakabe S, et al (1991) Atomic Data Nucl Data Tables 49(2):257. ISSN 0092-640X. https://www.sciencedirect.com/science/article/pii/0092640X91900272?via%3Dihub
112. Hashida M, et al (1996) Phys Rev A 54(5):4573. ISSN 1050-2947. https://doi.org/10.1103/PhysRevA.54.4573
113. Hashida M, et al (2011) Phys Rev A 83(3):032704. ISSN 1050-2947. https://doi.org/10.1103/PhysRevA.83.032704
114. Sakabe S, et al (1992) Phys Rev A 45(3):2086. ISSN 1050-2947. https://doi.org/10.1103/PhysRevA.45.2086
115. Pullins SH, et al (2000) Zeitschrift fur Physikalische Chemie 214(36770):1279. ISSN 09429352
116. Hause ML, et al (2013) J Appl Phys 113(16):163301. ISSN 0021-8979. https://doi.org/10.1063/1.4802432
117. Samaddar S, et al (2017) J Phys B: Atomic Mol Opt Phys 50(6):065202. ISSN 0953-4075. http://stacks.iop.org/0953-4075/50/i=6/a=065202?key=crossref.d43dbb80fec0cb3a0e2c71861c86f787
118. Ghanbari-Adivi E, et al. (2014) Phys Scripta 89(10):105402. ISSN 0031-8949. http://stacks.iop.org/1402-4896/89/i=10/a=105402?key=crossref.f4cb20c2206b4ade6932a9d9b53fc82d
119. Harris AL, et al (2010) Phys Rev A 82(2):022714. ISSN 1050-2947. https://doi.org/10.1103/PhysRevA.82.022714
120. Baer T, et al (1978) J Chem Phys 68(11):4901. https://doi.org/10.1063/1.435645
121. Yang W, et al (2018) Cit: Phys Plasmas 25(063521). https://doi.org/10.1063/1.5032276, http://aip.scitation.org/toc/php/25/6

122. Ryder C, et al (2015) Spectrochim Acta Part B: Atomic Spectrosc 113:16. ISSN 05848547.
     http://linkinghub.elsevier.com/retrieve/pii/S0584854715001962
123. Klose A, et al (2012) Nucl Instrum Methods Phys Res, Sect A: Accel, Spectrometers, Detect
     Assoc Equip 678(JUNE):114. ISSN 01689002

Chapter 3
Experimental Method Background

3.1 ISOLDE

3.1.1 Isotope Production

ISOLDE is an isotope separation 'online'[1] (ISOL) facility located at CERN. It was designed for the production, ionization, separation and transport of radioisotopes for a wide variety of research experiments [1, 2].

At ISOLDE, 1.4 GeV protons from the proton synchrotron booster (PSB) at CERN impinge upon a thick target which results in spallation, fragmentation and fission reactions with the isotopes of the target material. The isotope production yields critically depend on the target material used. Typical target materials include carbides (U, Si, Th, La), oxides (Mg, Ca), refractory metal foils (Ti, Ta, Nb) and molten metals (La, Pb, Sn), although over 100 different target materials have been tested to date [1]. Currents of 500–1000 A are used to heat 20 cm long tantalum tubes containing the material to temperatures between 600 and 2200 °C to enable fast release of the short-lived isotopes. The produced isotopes diffuse through the target material and effuse into an ion-source [3] to be ionized.

The use of thick targets allows measurable production yields of even very low cross-section exotic isotopes, however the diffusion time remains the primary limitation on the measurement of short-lived isotopes [4]. For the production of neutron-rich indium it was known that the highest yield would be found using UC_x target with a tungsten neutron converter [5] to suppress the nearby caesium mass contamination [6, 7]. A target temperature of around 1875 °C was found to be optimum for indium release from the target, much higher than 2200 °C can result in sintering of the target which reduces grain-boundary diffusion [8] and therefore reduces the yield of short-lived isotopes.

---

[1] 'online' refers to in-flight mass separation immediately following isotope production.

© The Editor(s) (if applicable) and The Author(s), under exclusive license to Springer Nature Switzerland AG 2020
A. R. Vernon, *Collinear Resonance Ionization Spectroscopy of Neutron-Rich Indium Isotopes*, Springer Theses,
https://doi.org/10.1007/978-3-030-54189-7_3

The yield of a given isotope available for measurement can be stated [2] as the product of the proton flux θ, cross section for the production reaction σ, number density of the target N and isotope-separation efficiency ϵ_{ISOL}

$$Y = \theta \sigma N \epsilon_{ISOL} \, , \tag{3.1}$$

where ϵ_{ISOL} depends on all efficiencies up to delivery to the central beamline

$$\epsilon_{ISOL} = \epsilon_{t_{1/2}} \epsilon_{diffusion} \epsilon_{effusion} \epsilon_{ionization} \epsilon_{separation} \epsilon_{transport} \, . \tag{3.2}$$

The coefficients ϵ give the efficiency for each extraction step following the production of the isotope of interest in the thick target to its transport to the CRIS beamline. These steps are further described in the sections below.

3.1.2 Ionization

The most common ion sources used with targets at ISOLDE are a surface ion source, plasma ion source or laser ion source. The most appropriate depends on the chemistry of the element to be ionized, surface and laser ionization can be complementary in many cases.

3.1.2.1 The Surface Ion Source

The most straightforward method of ionization from an ISOLDE target is by surface ionization. Figure 3.1 shows a schematic of the ISOLDE surface ion source in which atoms diffuse through a tantalum transfer line heated to $\sim$2200 °C [9]. It is possible

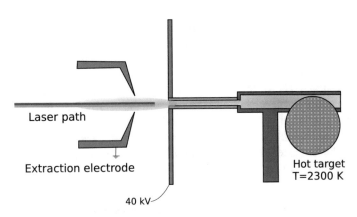

Fig. 3.1 Schematic of the surface ion source at ISOLDE with the laser ion source also in operation

for elements with an ionization potentials of less than ~6.2 eV to be efficiently surface ionized by collision with the heated surface in this way, with decreasing efficiency for higher ionization potentials [10]. Low ionization potential elements such as alkali metals constitute a large source of beam contamination.

If the ionization potential is too high for surface ionization or the element is volatile, then a plasma ion source configuration can be used [11]. The transfer line can also be cooled to reduce the diffusion of less volatile elements.

In the neuron-rich indium mass region (113 $\leqslant$ A $\leqslant$ 131), caesium was the most readily released contamination from the target and therefore the assumed largest source of collisional background at the CRIS beamline. An accident before the experiment to measure the neutron-rich indium isotopes led to venting of the target to atmosphere while still under heating, this would have affected the chemistry of the target and likely explains the high caesium contamination to indium ratios. The line temperature was reduced from 2000 °C to 1875 °C for these experiments to reduce the caesium contamination released from the target. This resulted in an increase in the ratio of indium to caesium yield from

$$\frac{^{128}\text{In}}{^{128}\text{Cs}} = \frac{1 \times 10^4 \ \mu\text{C}^{-1}}{7.4 \times 10^4 \ \mu\text{C}^{-1}} = 13.51\%,$$

to

$$\frac{^{128}\text{In}}{^{128}\text{Cs}} = \frac{1.9 \times 10^4 \ \mu\text{C}^{-1}}{5.5 \times 10^4 \ \mu\text{C}^{-1}} = 34.5\%,$$

as measured per μC of proton current by β^- spectroscopy on the ISOLDE tape station [1] (see also Table 3.1). A proton current of 2 μA was delivered from the PSB for the majority of experimental run time.

3.1.2.2 The Resonance Ionization Laser Ion Source (RILIS)

In most cases it is beneficial to use laser ionization in combination with the surface ion source. The resonance ionization laser ion source (RILIS) [12] is operated by overlapping lasers within the transfer line to resonantly excite and ionize the element of interest from the hot cavity. The ionization step can then proceed by non-resonant photoionization into the continuum, resonantly by excitation to an auto-ionizing state or by thermal ionization from a Rydberg state [13, 14].

As a resonant transition is used this technique can be highly selective, but this selectivity depends on the relative efficiency of laser ϵ_{laser} and surface ionization $\epsilon_{surface}$ to give a beam purity, P, as

$$P = \frac{\epsilon_{laser}}{\epsilon_{surface}}, \tag{3.3}$$

which heavily depends on the ionization potential of the isobar atoms.

Table 3.1 Comparison of indium yields to caesium contamination determined by switching the RILIS laser on and off. The enhancement was measured by a Faraday cup located after the HRS magnets, and additional with the ISOLDE tape station [1]. ISOLDE Database—[18, 19]

In mass	Beam current (pA)			Yield (ions/μC)		
	RILIS ON	RILIS OFF	Ratio	ISOLDE database	Tape. In	Tape. Cs
113	2	0.1	0.05	$>10^4$	–	–
114	2	0	0.00	$>10^4$	–	–
115	250	7	0.03	$>10^4$	–	–
116	6	0	0.00	$>10^4$	–	–
117	10	0.2	0.02	$>10^4$	–	–
119	15	0.6	0.04	$>10^4$	–	–
120	30	0.5	0.02	1.5×10^4	–	–
121	20	2	0.10	$>10^3$	–	–
122	30	0.5	0.02	$>10^3$	–	–
123	20	1	0.05	$>10^3$	–	–
124	13	0.1	0.01	$>10^3$	–	–
125	3	0	0.00	$>10^3$	–	–
126	2	0	0.00	$>10^3$	–	–
127	2	0.5	0.25	$>10^3$	–	–
128	0.5	0.2	0.40	4.4×10^3	1×10^4	7.4×10^4
129	1.2	0.5	0.42	1.1×10^3	–	–
130	2	1.75	0.88	$>10^2$	1×10^4	1×10^7
131	1.8	1.8	1.00	$>10^2$	–	–
132	3	3	1.00	8×10^3	–	5×10^8
133	3	3	1.00	9×10^2	–	–
134	–	–	–	95	–	–
135	–	–	–	2.4	–	–

The method shares much in common with experiments at the CRIS beamline, however the hot cavity environment causes significant Doppler broadening and complicates high-resolution laser spectroscopy. Figure 3.7 illustrates this point. In some heavy elements, with transitions which have large isotope shift and hyperfine constants, are typically resolvable with RILIS [15, 16].

The RILIS was used to enhance the indium production for the experiments of this work. The ionization scheme consisted of a 304 nm ($5s^2\,5p\,2P_{1/2} \rightarrow 5s^2\,5d\,2D_{3/2}$) step from a pulsed dye laser [17] and then a 532 nm step for non-resonant ionization, as shown in the scheme of Fig. 3.2.

As an example of the enhancement from using RILIS, at mass 115 the beam current increased from 12 pA to 500 pA by turning on the RILIS lasers, which should be almost entirely from ^{115}In. At higher masses the percentage of caesium contamination was large with RILIS lasers off, but much less with lasers on, with

Fig. 3.2 The indium ionization scheme used by RILIS during the measurements of radioactive indium isotopes. The 304 nm $(5s^2\,5p\,{}^2P_{1/2} \rightarrow 5s^2\,5d\,{}^2D_{3/2})$ transition gave the element selectivity

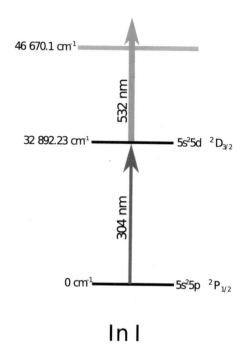

only a slight enhancement to the ionization of caesium from the non-resonant 532 nm step (due to heating the transfer line with 40 W of 532 nm light at 10 kHz).

The caesium contamination was significant enough to be the limiting factor in the measurements, as can be seen in Table 3.1. The RILIS ON and RILIS OFF measurements in this table refer to a remote shutter being open or closed which blocks the resonant transition in the scheme. This was after target temperature optimization from 1868 °C to 2000 °C, which increased the ^{130}In yield from 5×10^3 ions/μC to 1×10^4 ions/μC and ^{130}Cs yield from 3.3×10^6 ions/μC to 1×10^7 ions/μC, further decreasing the contamination ratio (the target temperature optimization was carried out before the previously mentioned line temperature optimization).

The suppression of the large caesium contamination was not enough to allow measurements further than ^{131}In at the time. The predicted yield of ^{130}Cs was 1×10^4 ions/μC from past experiments at ISOLDE [18, 19]. However yields are very element and target dependent [20, 21].

Further discussion of background suppression can be found in Sect. 6.3.3 of this thesis. Sufficient background suppression would allow further measurement of the neutron-rich indium isotopes using the CRIS setup despite increased contamination. Measurements with yields as low as 20 ions/s have been made with the CRIS setup in the past, on ^{78}Cu, when background was less of a limiting factor (0.0025 cps) [22], measurements up to ^{135}In are feasible in principle.

3.1.3 Online Isotope Separation

The choice of ion source provides a partial element specific selection, of the radioisotopes to be studied. This can be high element selectivity if the RILIS is used, or based only on the elements chemical properties for surface or plasma ion sources. The method of online isotope separation by magnets then allows a specific mass number to be measured by selecting out a specific A/q from the ionized beam. Two target areas and corresponding mass separating magnet arrangements are available at ISOLDE, the general-purpose separator (GPS) and the high-resolution separator (HRS).

The GPS consists of a single magnet, an electrostatic switchyard and movable plates which allows the delivery of up to three beams to different experiments simultaneously, with a maximum of ±15 A/q difference from the central mass [23]. The HRS is an array of two magnets and ion optics to produce a higher mass resolving power, with a designed mass resolution of greater than 5000, in theory [2]. The caesium contamination during this experiment significantly reduced the selectivity and mass resolving power however. Beam energies of up to 50 keV can be reached after ionization and the HRS beamline. An energy of 40 keV was used for the indium measurements presented in this thesis. This energy was found to give the best transmission to and through the CRIS beamline and ensured reduced Doppler broadening (Fig. 3.3).

After separation only electrostatic elements are used for ion transport at ISOLDE, making the delivery to experiments independent of mass [1] (apart from the ISCOOL for which the RF frequency is changed according to the mass). Both separators can deliver beams to experiments in the hall through a merging switchyard to the central beamline. At present experiments at the CRIS beamline can only be performed with the HRS due to the requirement of the ion cooler-buncher ISCOOL (see Sect. 3.1.4) which is located on the HRS branch.

3.1.4 The ISCOOL Cooler Buncher

A principle requirement for the CRIS technique to be efficient with the pulsed non-resonant final ionization step lasers which are used, is the use of bunched atomic beams. An additional advantage is that the bunching duty cycle allows a proportion of background counts to be gated out. The ISOLDE cooler-buncher ISCOOL [24, 25] is used for this purpose. The trap consists of a radio-frequency quadrupole (RFQ) which decelerates ions and reduces their energy spread using a buffer gas. The device is shown schematically in Fig. 3.4. A linear DC voltage is superimposed with the oscillating electric field in order to guide the ions through the cooler, a fast release switch drops the voltage on the end plate of the cooler in less than 100 ns [26] to release the bunch.

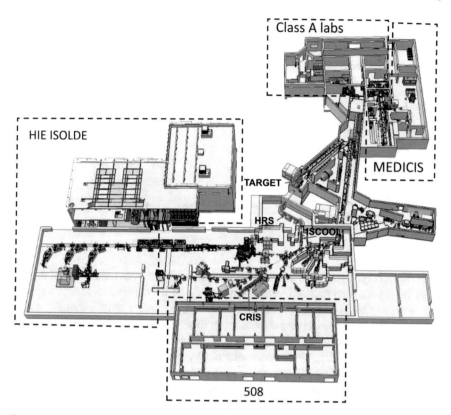

Fig. 3.3 Layout of the ISOLDE facility at CERN. Reprinted with permission from Catherall, R. et al. The ISOLDE facility. J. Phys. G Nucl. Part. Phys. 44, (2017). Copyright 2020 by the Institute of Physics

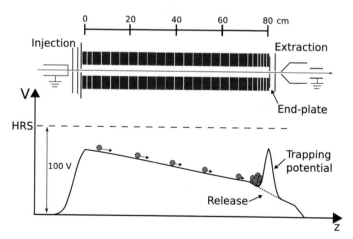

Fig. 3.4 Schematic of the segmented radio frequency quadrupole trap ISCOOL. Adapted from [24]

Fig. 3.5 An example time-of-flight spectrum of the ion bunch released from ISCOOL (10 ms bunch time), for ^{115}In. T_0 is the laser release time i.e. the time of flight of the ions is from the CRIS interaction region to the detector

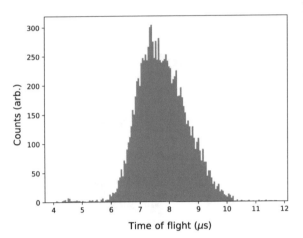

Helium is typically used as a buffer gas due to its light mass, the mass of the gas must be lighter than that of the incident ions (m_B) for cooling to take place [27], $m_B > m_{He}$. This is one limitation in the bunching of lighter mass isotopes such as carbon as the increase in required bunching time and the increased reactivity with contaminants in the buffer gas significantly reduces the bunching efficiency. Once extracted from the trap, electrostatic elements are used to guide the bunch to the CRIS beamline where the beam emittance should remain constant [28].

A temporal spread of $<5\,\mu s$ is required for the bunch to fit within the CRIS interaction region at 40 keV. For these measurements of indium, the average time of flight from the cooler was 80 μs with a temporal spread of 2 μs. See Fig. 3.5 for an example ^{115}In time-of-flight spectrum of the ion bunch released from ISCOOL.

An upper limit for the energy spread of the bunches can be given as 0.5 eV, based on the 40(10) MHz Gaussian linewidth of resonances obtained with the high-resolution laser used for these experiments.

3.2 The Collinear Resonance Ionization Spectroscopy (CRIS) Experiment

The collinear resonance ionization spectroscopy (CRIS) technique has been developed for the measurement of short-lived exotic isotopes, with the two principle advantages over conventional laser-spectroscopy methods of increased resolution and sensitivity, as compared to in-source laser spectroscopy [29] and collinear fluorescence detection laser spectroscopy [30] respectively. Figure 3.6 shows the layout of the CRIS beamline at CERN-ISOLDE.

Accelerating the beam to energies greater than $\sim$5 keV causes the Doppler broadening observed in measured atomic transitions to be reduced significantly [31, 32], where the laser linewidth of $\sim$34 MHz [33] was the primary limitation in resolution

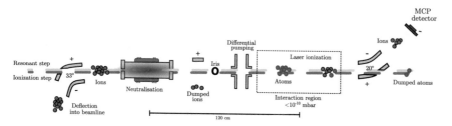

Fig. 3.6 Schematic of the collinear resonance ionization spectroscopy beam line at CERN-ISOLDE used to make the measurements of this work

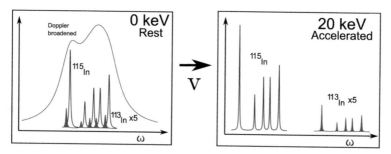

Fig. 3.7 Idealised hyperfine spectra of stable indium isotopes ^{115}In and ^{113}In to illustrate the advantage of collinear laser spectroscopy over in-source laser spectroscopy in a simple way

during these experiments. The reduction in velocity spread of the beam, δv, with an increase in beam energy, ΔT_B, is due to conservation of energy as

$$\Delta T_B \approx m_B v \delta v . \tag{3.4}$$

Figure 3.7 shows a simulated comparison to a Doppler broadened spectrum typical of in-source laser spectroscopy. Large linear stopping powers exist for ions interacting with matter due to the Coulomb interaction, compared to the smaller cross section for interaction of photons via the photoelectric effect for the same detection volume. In addition, the charge of ions allow them to be electro-statically guided into a particle detector, removing solid angle losses as compared to the detection of photons which necessitate lenses and angular arrays for efficient collection [34]. Ionization of radioisotopes therefore allows for improved detection efficiency ($\sim$100% [35]) compared to conventional fluorescence photon detection by photomultiplier tubes (<30% [36]). However, the total signal rate for a measurement is not only a function of detection efficiency but also of the total efficiency of the experimental setup and also the atomic transition strength in the case of photon detection. The use of photomultiplier tubes has the additional disadvantage of higher background rates compared to ion detectors, from dark counts and scattered light.

A pair of 33° electrostatic bending plates allows overlap of the probe lasers with the beam of ions deflected from ISOLDE. The ions are then directed through a charge-exchange cell (see Sects. 3.2.1 and 2.4) to neutralise by electron pick up reactions with an alkali metal vapour.

The ionisation of neutralised atoms often requires several laser-excitation steps to exceed the ionization potential of the element. The laser beams are overlapped in a collinear geometry in the interaction region of CRIS with the neutralised atoms. The frequency of a resonant step laser is scanned to excite the transitions between the hyperfine levels of the isotope. Typically, a final non-resonant high-power laser is used to ionize the atoms once they have been resonantly excited by a first-step transition. The use of multiple excitation steps additionally allows for highly selective laser ionization and detection of very low production yield isotopes, even in the presence of beams with substantial isobaric contamination.

The selectivity, S, for ionization with n resonant laser excitation steps is given by

$$S = \prod_{n=1}^{N} \left(\frac{\delta\omega_n}{\Gamma_n} \right)^2 = \prod_{n=1}^{N} S_n , \tag{3.5}$$

where $\delta\omega_n$ is the separation of transitions from other states and Γ_n is the FWHM of the transitions [26].

Collisional re-ionization of the neutral beam or non-resonant photoionization, in the same time window as the resonant laser ionization, are presently the primary sources of background for the technique. To reduce the collisional background the 'interaction region' of the CRIS beamline is kept under ultra-high vacuum conditions of $<10^{-9}$ mbar. Typically laser ionization schemes are chosen so that the penultimate laser step is close to the IP of the element, so that the final non-resonant step can be performed with a long wavelength step (e.g. 1064 nm light where possible), to reduce photoionization background. Mass contamination from isobars of the isotope under study is typically the largest source of background. The contaminant contribution to the final atom bunch can be magnitudes larger than the element of interest in some cases, and can therefore prevent or slow down measurement due to the background count rate.

MagneToF [37] or micro-channel plate [38] (MCP) detectors are used to detect ions directly, by being placed in the path of the beam following ionization and deflection through a 20° bend.

The ions can alternatively be detected indirectly via secondary electrons emitted by impinging the ion beam onto a conversion dynode. The electrons can then detected by an MCP detector, this arrangement is typically used to avoid implanting long-lived contamination into the detector.

Selective ionization also has the additional advantage that the ion can be used for decay spectroscopy in a laser assisted nuclear decay spectroscopy arrangement, 'LANDS' [26, 39], allowing measurements to be made on individual ground or isomeric states of an isotope [40, 41].

In cases where isobaric contamination creates an overwhelming background and the half-life of the isotope is sufficiently short then decay assisted laser spectroscopy, 'DALS', can be used, in which the laser frequency is scanned and the decay spec-

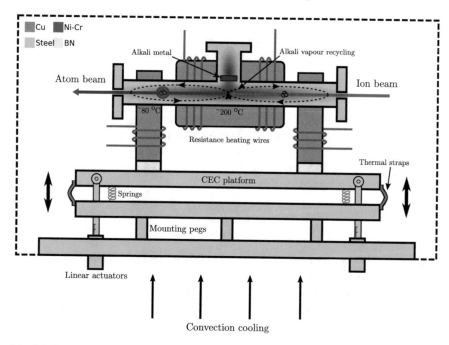

Fig. 3.8 Schematic of the design of the charge-exchange cell mounted in the CRIS beamline. The BN (Boron Nitride) spacers will electrical isolation of the cell, Cu spacers are presently used in their place, and cartridge heaters now replace the Ni-Cr resistance heating wire

troscopy events used to create the hyperfine spectrum, resulting in greatly improved background suppression [41].

3.2.1 The Charge-Exchange Cell

The laser ionization schemes and ion optical arrangements needed are greatly simplified by performing CRIS on the atomic form of the radioactive isotopes delivered from ISOLDE, rather than on the 1^+ ion directly. For this purpose ions are first passed through a charge-exchange cell filled with alkali metal vapour. The design adopted at the CRIS beamline is a horizontal charge-exchange cell with cooled ends. A steel mesh inside the cell allows condensation and recirculation of the metal vapour, increasing the effective lifetime of the alkali metal reservoir [42, 43] and reducing the deposition rate of material outside of the cell. A schematic of the cell is shown in Fig. 3.8. An adjustable 5-axis mount was designed for optimization of beam transport through the cell in-situ under vacuum.

Using the attenuation model [44], the total efficiency for neutralisation an ion passing through an effective path length, L, of a vapour composed of the free atom B, of atom number density n, can be estimated by

$$\epsilon_{CE} = 1 - \exp(-n\sigma_{CE}L); \tag{3.6}$$

Both n and L are dependent upon the dimensions of the vapour cell used [42, 43, 45], but can be estimated from temperature measurements using empirical vapour-pressure relations [46]. The effective length, L, can roughly be taken as the cell inner length of 15 cm. The number density, n, can be estimated assuming a saturated vapour pressure at temperature T K from

$$n = \frac{10^{(A-B/T)} N_A}{V_{mol}\, p_0} \tag{3.7}$$

where N_A is Avogadro's number, V_{mol} is the standard molar volume, p_0 is the standard pressure and A and B are empirical constants for sodium with values of 4.704 and 5377 respectively [46]. The total neutralisation cross section, σ_{CE}, can vary significantly depending on the incident ion to be neutralised and alkali metal used and is a function of many individual atomic states in reality.

Prior to the installation of the new mounting system, thermal simulations were performed for the ultra-high vacuum chamber and charge-exchange cell. The results of a few key cases are shown in Fig. 3.9b). Ultimately a passive cooling design was possible to produce potassium vapour provided that either the outer chamber was cooled, or that sufficient thermal conductance was added to the design by through the mounting geometry, materials or thermal straps. A temperature surface plot of the final design is shown in Fig. 3.9a where six 4 cm wide copper braids were used as thermal straps and steel pegs were used to mount the cell. When loaded with potassium and heated to a central temperature of $T_{cent} = 214\,°C$, the stands (the blocks raising the cell to the height of the ion beam) of the cell reach $T_{ends} = 82\,°C$ when left to reach thermal equilibrium. These values are between values simulated for the vacuum chamber actively cooled to 20 °C and left to radiately cool to the room. The recirculation of the potassium was not included in the simulations, therefore the temperature profiles are an underestimate of those expected in practice, which may explain the discrepancy. The design has operated for $>200\,h$ with the passively cooled configuration. As the construction of the cell allows for heating up to 450 °C, air cooling has since been incorporated into the stands of the cell to allow for the use of sodium at temperatures of $T_{cent} = 264\,°C$ and $T_{ends} = 113\,°C$. Other low melting point elements may be desirable for neutralisation in the future, as the total neutralisation efficiency σ_{CE} and final atomic populations of the atom product are element specific for both the atom and ion reactants (Sect. 2.4).

The CRIS interaction region is located 120 cm downstream of the CEC, at nominal pressures of 3×10^{-10} mbar compared to the CEC region at 5×10^{-8} mbar with estimated pressures of the order 1×10^{-4} mbar inside the cell [46]. Although differential pumping apertures are installed along the beamline to allow for the large pressure

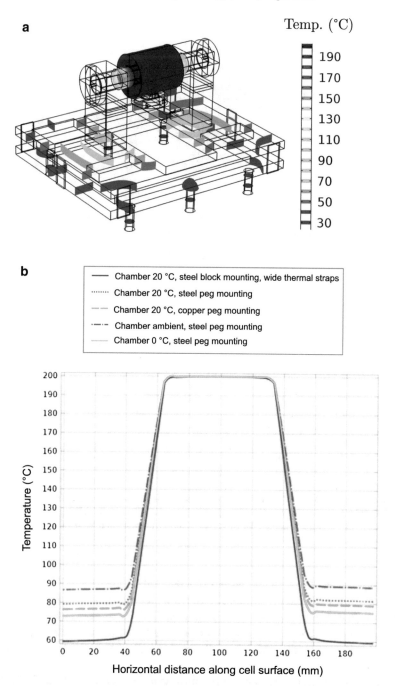

Fig. 3.9 a Surface plot of simulated temperature for the final design used for the charge-exchange cell, using steel peg mounting. **b** Horizontal temperature profiles along the charge-exchange cell outer surface simulated under different mounting and chamber boundary temperatures. Simulations were performed using COMSOL Multiphysics®

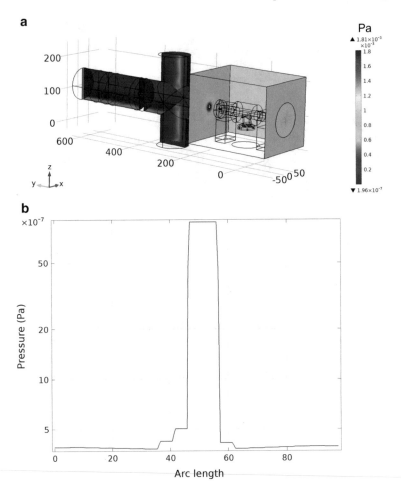

Fig. 3.10 **a** Surface plot of the pressure gradient caused by molecular beaming from the cell under free molecular flow conditions. **b** Pressure gradient due to molecular beaming into the interaction region of CRIS, along the arc shown in red in **a** Simulations were performed using COMSOL Multiphysics®

differences, as the CEC vapour has to share the same collinear axis as the laser path and interaction region there is a higher probability for the 'molecular beaming' effect [47–49]. That is neutral atoms or molecules not belonging to the ion beam passed through the cell travelling downstream due to their thermal velocities, creating an effectively higher density region for collisional re-ionization and therefore increased background.

The assumption of free-molecular flow can be made under high vacuum ($<1\times10^{-3}$ mbar [46]), although the flow regime inside the cell will be transient, the free-molecular flow assumption allowed the simulations shown in Fig. 3.10a, b to be performed. The simulated effective pressure in the interaction region from the molecular beaming effect is of the order 1×10^{-7} mbar. This will contribute to the collisional background for these experiments, but will depend on the atomic species entering the interaction region (discussed in Sect. 6.3.3).

The horizontal charge-exchange cell design was used as it is known to be a reliable method of neutralisation over a long running period. However other arrangements have been implemented, such as a vertical configuration, which avoids oxidation of the alkali metal if the beamline is vented [45]. Ablation of the alkali metal has also been found to be as efficient as vapour cells for neutralising due to a high instantaneous atom density [50], but instabilities between laser pulses can be large, especially when using a solid target. Thin carbon foil arrangements can also be very efficient for neutralisation [51] but are incompatible with the collinear pulsed laser geometry of CRIS. Negative ion production can be large using foils, a fraction of negative ions can also be found using vapour cells [52], depending on the operating temperature. The volatile molecule isobutane allows for a high vapor pressure and is very effective for negative ion production [53].

References

1. Catherall R et al (2017) J Phys G: Nucl Part Phys 44(9):094002. ISSN 0954-3899. http://stacks.iop.org/0954-3899/44/i=9/a=094002?key=crossref.cb9ee490ac00453f72e007300938b2c6
2. Jonson B et al (2000) Hyperfine Interact 129(1/4):1. ISSN 03043834. http://link.springer.com/10.1023/A:1012689128103
3. Lettry J et al (1997) Nucl Instrum Methods Phys Res Sect B: Beam Interact Mater At 126(1–4):170. ISSN 0168-583X. https://www.sciencedirect.com/science/article/pii/S0168583X96010889
4. Van Duppen P (2006) Isotope separation on line and post acceleration. Springer, Berlin. http://link.springer.com/10.1007/3-540-33787-3_2
5. Gottberg A et al (2014) Nucl Instrum Methods Phys Res Sect B: Beam Interact Mater At 336:143. ISSN 0168-583X. https://www.sciencedirect.com/science/article/pii/S0168583X1400528X, https://www.sciencedirect.com/science/article/pii/S0168583X1400528X#f0010
6. Dillmann I et al (2002) Eur Phys J A 13(3):281. ISSN 14346001
7. Köster U (2002) Eur Phys J A 15(1–2):255. ISSN 14346001. http://link.springer.com/10.1140/epja/i2001-10264-2
8. Bhadeshia HKDH et al (2017) Steels: microstructure and properties, 4th edn. Butterworth-Heinemann, Oxford
9. Goodacre TD et al (2016) Nucl Instrum Methods Phys Res Sect B: Beam Interact Mater At 376:39. ISSN 0168-583X. https://www.sciencedirect.com/science/article/pii/S0168583X16002111
10. Kirchner R (1981) Nucl Instrum Methods Phys Res 186(1–2):275. ISSN 0167-5087. https://www.sciencedirect.com/science/article/pii/0029554X81909162
11. Stora T (2013) Radioactive ion sources. Technical report, CERN. https://cds.cern.ch/record/1693046/files/p331.pdf
12. Marsh BA et al (2013) Nucl Instrum Methods Phys Res Sect B: Beam Interact Mater At 317(PART B). ISSN 0168583X
13. Rothe S et al (2016) Nucl Instrum Methods Phys Res Sect B: Beam Interact Mater At 376:91. ISSN 0168583X. http://dx.doi.org/10.1016/j.nimb.2016.02.024

14. Jading Y et al (1997) Nucl Instrum Methods Phys Res Sect B: Beam Interact Mater At 126(1–4):76. ISSN 0168-583X. https://www.sciencedirect.com/science/article/pii/S0168583X9601018X

15. Marsh BA et al (2018) Nat Phys 1. ISSN 1745-2473. http://www.nature.com/articles/s41567-018-0292-8

16. Seliverstov M et al (2013) Phys Lett B 719(4–5):362. ISSN 0370-2693. https://www.sciencedirect.com/science/article/pii/S0370269313000841?via%3Dihub#fg0040

17. Fedosseev VN et al (2012) Physica Scripta 85(5):058104. ISSN 0031-8949. http://stacks.iop.org/1402-4896/85/i=5/a=058104?key=crossref.c4b819f0a5bcf527b8fa8270eb7e4ee1

18. Koster U et al (2008) Nucl Instrum Methods Phys Res Sect B: Beam Interact Mater At 266(19-20):4229. ISSN 0168-583X. https://www.sciencedirect.com/science/article/pii/S0168583X08007118

19. Johnston K, Stora T, Turrion M, Herman-Izycka U (2016) ISOLDE yield database. http://test-isolde-yields.web.cern.ch/test-isolde-yields/query_tgt.htm

20. Beyer GJ et al (2003) Nucl Instrum Methods Phys Res B 204:225. www.elsevier.com/locate/nimb

21. Bergmann U et al (2003) Nucl Instrum Methods Phys Res B. www.elsevier.com/locate/nimb

22. De Groote R et al (2017) Phys Rev C 96(4):1. ISSN 24699993

23. Kugler E (2000) Hyperfine Interact 129(1/4):23. ISSN 03043834. http://link.springer.com/10.1023/A:1012603025802

24. Mané E et al (2009) Eur Phys J A 42(3):503. ISSN 14346001

25. Jokinen A et al (2003) Nucl Instrum Methods Phys Res Sect B: Beam Interact Mater At 204:86. ISSN 0168-583X. https://www.sciencedirect.com/science/article/pii/S0168583X02018943

26. Lynch KM (2013) PhD thesis, School of Physics and Astronomy, The University of Manchester, Manchester, United Kingdom

27. Nieminen A et al (2000) Hyperfine Interact 127:507. https://link.springer.com/content/pdf/10.1023%2FA%3A1012694407204.pdf

28. Müller-Kirsten HJW (2013) Basics of statistical physics. World Scientific, Singapore. https://www.worldscientific.com/worldscibooks/10.1142/8709

29. Fedosseev V et al (2017) J Phys G: Nucl Part Phys 44(8):084006. ISSN 0954-3899. http://stacks.iop.org/0954-3899/44/i=8/a=084006?key=crossref.d6d5210c0d6f03d7db22808c72e24c0f

30. Anton KR et al (1978) Phys Rev Lett 40(10):642. ISSN 0031-9007. https://link.aps.org/doi/10.1103/PhysRevLett.40.642

31. Wing WH et al (1976) Phys Rev Lett 36(25):1488. ISSN 0031-9007. https://link.aps.org/doi/10.1103/PhysRevLett.36.1488

32. Kaufman S (1976) Opt Commun 17(3):309. ISSN 00304018. http://linkinghub.elsevier.com/retrieve/pii/0030401876902674

33. Sonnenschein V (2015) PhD thesis, Department of Physics, University of Jyvaskyla

34. Knoll GF (2010) Radiation detection and measurement. Wiley, Hoboken

35. Gilmore IS et al (2000) Ion detection efficiency of microchannel plates. Technical report, National Physical Laboratory. http://resource.npl.co.uk/docs/science_technology/nanotechnology/nanopubs/iondetectionefficiency.pdf

36. Hamamatsu (2007) Photomultiplier tubes photon is our business basics and applications, 3rd edn (Edition 3a). Technical report, Hamamatsu. https://www.hamamatsu.com/resources/pdf/etd/PMT_handbook_v3aE.pdf

37. ETP (2018) ETP ion detect. https://www.etp-ms.com/

38. Liénard E et al (2005) Nucl Instrum Methods Phys Res Sect A: Accel Spectrom Detect Assoc Equip 551(2-3):375. ISSN 0168-9002. https://www.sciencedirect.com/science/article/pii/S0168900205013811

39. Lynch K et al (2017) Nucl Instrum Methods Phys Res Sect A: Accel Spectrom Detect Assoc Equip 844:14. ISSN 01689002

40. Lynch KM et al (2016) Phys Rev C 93(1):014319. ISSN 2469-9985. http://link.aps.org/doi/10.1103/PhysRevC.93.014319

41. Lynch KM et al (2014) Phys Rev X 4(1):1. ISSN 21603308. http://link.aps.org/doi/10.1103/PhysRevX.4.011055
42. Paulo S (1981) Nucl Instrum Methods 179:195
43. Hiddleston HR (1977) IEEE Trans Nucl Sci NS-24(3):1600
44. Hasted JB (1972) Physics of atomic collisions, 2d edn. American Elsevier. https://bibdata.princeton.edu/bibliographic/1917772
45. Klose A et al (2012) Nucl Instrum Methods Phys Res Sect A: Accel Spectrom Detect Assoc Equip 678:114. ISSN 01689002
46. Haynes WM (2015) CRC handbook of chemistry and physics: a ready-reference book of chemical and physical data, 96th edn. CRC Press, Boulder
47. Zhang S et al (2012) Phys Procedia 32:513. ISSN 1875-3892. https://www.sciencedirect.com/science/article/pii/S1875389212010127
48. Shiwei Z et al (2008) In: 2008 international conference on computer science and information technology, pp 486–491. IEEE. http://ieeexplore.ieee.org/document/4624916/
49. Szwemin P et al (1993) Vacuum 44(5–7):451. ISSN 0042-207X. https://www.sciencedirect.com/science/article/pii/0042207X9390070Q?via%3Dihub
50. Kubota M et al (1999) Nucl Instrum Methods Phys Res Sect B: Beam Interact Mater At 149:514
51. Bürgi A et al (1993) J Appl Phys 73(9):4130. ISSN 0021-8979. http://aip.scitation.org/doi/10.1063/1.352846
52. Heinemeier J et al (1978) Nucl Instrum Methods 148(3):65. ISSN 0029554X
53. Freeman SP et al (2015) Nucl Instrum Methods Phys Res Sect B: Beam Interact Mater At 361:229. ISSN 0168-583X. https://www.sciencedirect.com/science/article/pii/S0168583X15003821
54. CERN-ISOLDE (2018) Ion source | RILIS at ISOLDE. http://rilis.web.cern.ch/ion-source

Chapter 4
Laser Spectroscopy Setup for the Indium Experiments

A combination of pulsed and continuous-wave (CW) lasers were required to perform the measurements described in this thesis. The high-resolution measurements were made using an injection-seeded pulsed titanium:sapphire (Ti:Sa) laser (Sect. 4.1.4) seeded using a tunable M-Squared SolsTiS CW Ti:Sa laser (Sect. 4.1.6). This provided the light for the 246.8 nm ($5p\,^2P_{3/2} \rightarrow 9s\,^2S_{1/2}$), 246.0 nm ($5p\,^2P_{1/2} \rightarrow 8s\,^2S_{1/2}$), 283.7 nm ($5p\,^2P_{3/2} \rightarrow 5s5p^2\,^4P_{5/2}$) and 410 nm ($5p\,^2P_{1/2} \rightarrow 6s\,^2S_{1/2}$) transitions. The ionization schemes are indicated in Fig. 4.1. These four transitions were used for the offline measurements of naturally abundant ^{113,115}In isotopes produced with an ablation ion source, detailed in Chap. 5.

A Spectron Spectrolase 4000 pulsed dye laser (PDL) (Sect. 4.1.1) and a pulsed 'Z cavity' Ti:Sa laser [1] (Sect. 4.1.5) were used to search for initial resonance signals before optimisation. Both are broadband (BB) lasers, their large spectral linewidth enabled faster searching of frequency space to identify resonant signals. These BB lasers could then also be used for scanning of lower production yield indium isotopes if there was insufficient experimental time remaining for high-resolution measurements. The PDL was additionally used as a second step laser for the $5p\,^2P_{1/2} \rightarrow$ $6s\,^2S_{1/2}$ (410 nm) + $6s\,^2S_{1/2} \rightarrow 5s^28p\,^2P_{3/2}$ (571 nm) scheme. A Litron twin-head Nd:YAG pulsed laser (Sect. 4.1.2) with higher harmonic generation was used to provide the 1064 nm and 532 nm light for the non-resonant ionization steps. The $5p\,^2P_{3/2}$ $\rightarrow 9s\,^2S_{1/2}$ (246.8 nm) and $5p\,^2P_{1/2} \rightarrow 8s\,^2S_{1/2}$ (246.0 nm) transitions were used for the online measurements of neutron-rich indium isotopes.

Table 4.1 summarises the lasers required for these experiments and the purpose of each transition used. Figure 4.3 gives a convenient overview of the different combinations of Ti:Sa laser, dye laser, and higher harmonics used to access wavelengths in the range 200–1000 nm in modern laser spectroscopy experiments.

The layout of the lasers for the experiments is shown in Fig. 4.2. A combination of polarising beamsplitter cubes and 90° flip mounted mirrors allowed fast switching between each of these schemes (Fig. 4.3).

© The Editor(s) (if applicable) and The Author(s), under exclusive license
to Springer Nature Switzerland AG 2020
A. R. Vernon, *Collinear Resonance Ionization Spectroscopy
of Neutron-Rich Indium Isotopes*, Springer Theses,
https://doi.org/10.1007/978-3-030-54189-7_4

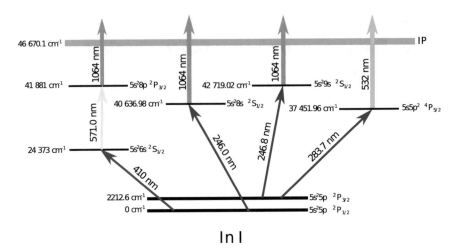

Fig. 4.1 The multi-step laser ionization schemes used in this work. See Table 4.1 for the purpose of each

Table 4.1 Summary of the transitions used to measure hyperfine structures in this work and their specific purpose. See Fig. 4.1 for the schemes

Lower state	Upper state	Transition	Fundamental	Laser	Distinct purpose
$5p\,^2P_{3/2}$	$9s\,^2S_{1/2}$	246.8 nm	740.4 nm	IS	$I^\dagger$, Q_S sensitivity
		246.8 nm	740.4 nm	Z cavity	BB search
$5p\,^2P_{1/2}$	$8s\,^2S_{1/2}$	246.0 nm	738.0 nm	IS	μ sensitivity
$5p\,^2P_{3/2}$	$5s5p^2\,^4P_{5/2}$	283.7 nm	851.1 nm	IS	$I^\dagger$, μ, Q_S sensitivity
		283.7 nm	566 nm	PDL	BB search
$5p\,^2P_{1/2}$	$6s\,^2S_{1/2}$	410 nm	820 nm	IS	s-orbit calculations
$6s\,^2S_{1/2}$	$5s^28p\,^2P_{3/2}$		571 nm	PDL	Followed 410.0 nm step
Upper states	Continuum		1064 nm & 532 nm	Litron	Non-resonant ionization

†—spin determination not possible for $I > 4$
IS—Injection seeded
PDL—Pulsed Dye Laser

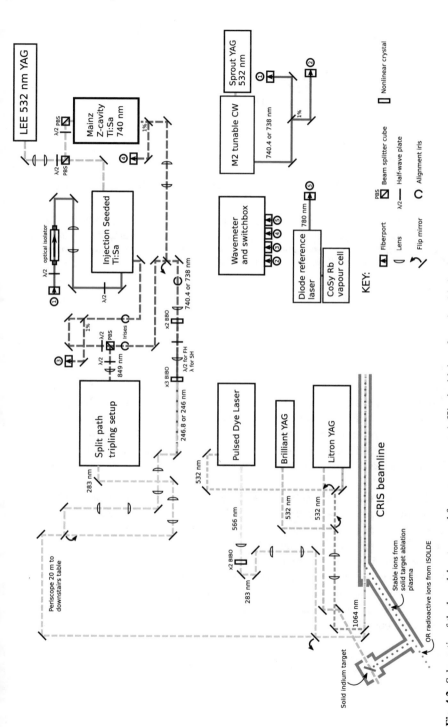

Fig. 4.2 Schematic of the laser lab setup used for measurements 'offline' with ions from the ablation ion source as well as 'online' with radioactive beam. 2x BBO and 3x BIBO indicate the doubling and tripling crystals used in the linear tripling arrangement

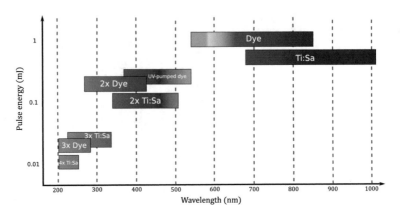

Fig. 4.3 Overview of typical accessible wavelengths with the combination of Ti:Sa and dye lasers with higher harmonic generation (2x etc.) crystals. The given order of magnitude of pulse energies are indicated, using values from a range of commonly used dyes and with a high repetition rate 532 nm, or 355 nm (for 'UV-pumped' dyes) Nd:YAG pump lasers [2]

4.1   Lasers

4.1.1   Pulsed Dye Laser (PDL)

The Spectron Spectrolase 4000 pulsed dye laser was used to provide broadband 567 nm light, to be frequency doubled for the 5p ^{2}P$_{3/2}$ → 5s5p^2 ^{4}P$_{5/2}$ (283.7 nm) transition. The 283.7 nm ionization scheme is shown in Fig. 4.1 and the schematic for the setup is shown in Fig. 4.2. This laser has a linewidth of 14 GHz in the case of frequency-doubled light. Frequency-doubled light from this laser was used for the indium spectra collected using the ablation ion source, shown by the red spectrum in Fig. 4.4.

This laser was additionally used to provide the fundamental light at 571 nm for the second step of the 410 nm ionization scheme. Broadband lasers are often used for this purpose, or to excite to auto-ionizing states, as their broad linewidth can often cover the width of the hyperfine splitting for a state.

The laser dye used for both of these transitions was Rhodamine 6G, which has a fluorescence peak centred at 566 nm when dissolved in ethanol [3]. One of the heads of the twin Litron Nd:YAG laser was used as a pump laser for the dye. A laser pulse energy of 0.7 mJ was measured for the fundamental light produced using this dye and pump combination.

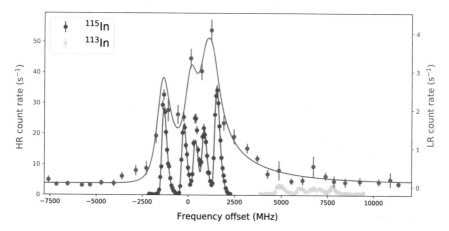

Fig. 4.4 Hyperfine spectra of atomic [115]In measured with the 283 nm transition under different conditions. (Red) Linewidth of 14 GHz. Low-resolution scan using the Spectron pulsed dye laser with indium ions created from the laser ablation ion source. (Blue) Linewidth of 280(60) MHz. High-resolution scan with the injection-seeded Ti:Sa laser with indium ions created from the laser ablation ion source, before energy spread reduction. The lower atomic abundance [113]In can be seen in light blue, but requires high resolution to separate the isotope from [115]In

4.1.2 Litron Twin-Head Nd:YAG Pulsed Laser

The Litron LPY 601 50-100 PIV laser was used to provide the final non-resonant ionization step in all of the schemes used for the measurements (Fig. 4.1).

The laser uses Nd:YAG rods as a lasing medium to produce 1064 nm light (~80 mJ) or 532 nm light (~50 mJ), following a second-harmonic-generation crystal. T100 Hz repetition rate of this laser was the limitation to the duty cycle of the other laser steps and ion bunch release cycle during the experiments.

As the laser allows independent operation of its heads, one head could remain in operation at 1064 nm to provide the non-resonant ionization step whilst the other could be used with a slight offset in time for the laser ablation ion source. A pulse duration of 10 ns is specified for this laser. A KD*P (potassium dideuterium phosphate) type Pockels cell is used for Q-switching in this laser, they have opening times of typically 1 ns [4].

In the configuration where 532 nm light from one of the heads was also used to pump the PDL laser, an additional Nd:YAG laser was required. The Quantel Brilliant Nd:YAG laser was used to provide the 1064 nm non-resonant step light (0.85 J) for this purpose, however this laser had a limited repetition rate 10 Hz (Fig. 4.2).

4.1.3   LEE Nd:YAG Pulsed Laser

The LDP-100MQ LEE laser is a diode pumped Nd:YAG laser. It was used in its second harmonic generation configuration to provide 532 nm pulsed pump light to both the injection-seeded and Z-cavity Ti:Sa lasers. This was implemented by polarising beam-splitter cubes and half-wave plates to distribute the power between the cavities as needed, see Fig. 4.2. The laser was operated at 1 kHz (10 kHz operation is possible) during the experiments, providing up to 10 mJ/pulse during the online laser spectroscopy (which was sufficient to simultaneously pump two Z-cavity lasers and a injection-seeded Ti:Sa laser). During the measurements of neutron-rich indium, the LEE laser served as the master trigger for the digital-delay pulse generator (Sect. 4.5), which synchronised the other lasers and ion release signals. The Q-switch of this laser has an internal compensation for the variable power from the diode pump. Therefore although external triggering of this laser is possible, it would cause a significant laser pulse jitter relative to the other lasers.

4.1.4   Injection-Seeded Pulsed Ti:Sapphire Laser

An injection-seeded Ti:Sapphire laser was used as the principal laser to provide narrow linewidth, pulsed laser light for the 5p $^2P_{3/2}$ → 9s $^2S_{1/2}$ (246.8 nm), 5p $^2P_{1/2}$ → 8s $^2S_{1/2}$ (246.0 nm), 5p $^2P_{3/2}$ → 5s5p^2 $^4P_{5/2}$ (283.7 nm) and 5p $^2P_{1/2}$ → 6s $^2S_{1/2}$ (410 nm) transitions in this work. The cavity itself produces fundamental light of 750–850 nm (mostly a limitation of the mirror set used [5, 6]). The light was frequency tripled to produce the UV light for the 246.8 nm, 246.0 nm and 283.7 nm transitions used in these experiments.

The purpose of injection-locking [8] in this context is to use a 'master' or 'seed' laser with a very narrow linewidth to provide amplified pulsed light with a much narrower linewidth than an unseeded Ti:Sa laser. The CW M-Squared laser was used as the seed laser for this work. The injection-locked cavity is then said to be the 'slave' laser. A schematic layout of the laser setup with the required locking electronics is shown in Fig. 4.5.

The particular injection-seeded Ti:Sa cavity used for these experiments follows the bow-tie design laid out by [7]. The construction and characterization of this cavity is detailed in [5, 6].

The locking feedback loop was implemented using a piezo-actuated mirror and cavity light measured on a photodiode via a coupling mirror. The mode of the cavity can then be locked to a mode generated with the seed light.

By overlapping the seed light with pulsed-laser light, pumping a Ti:Sa crystal via an input coupler, amplification can be achieved with a spectral bandwidth of 20 MHz [6]. Typically only a few mW of seed light is needed.

The pump laser used for the injection-seeded Ti:Sa case was the LEE Nd:YAG laser, which delivered 1.2–1.4 mJ of 532 nm light to produce 160 μJ of broadband

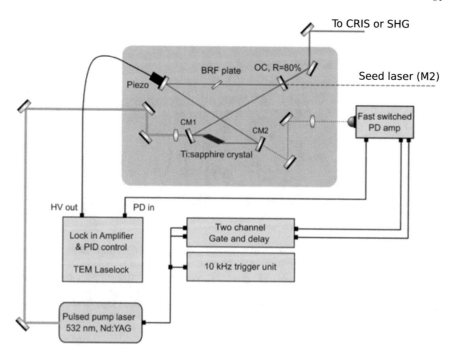

Fig. 4.5 The layout of the injection-seeded bow-tie Ti:Sa laser with the locking electronics, designed by [7]. Figure modified from [5] (Reprinted with permission from Sonnenschein, V. et al. Characterization of a pulsed injection-locked Ti:sapphire laser and its application to high resolution resonance ionization spectroscopy of copper. Laser Phys. 27, 085701 (2017). Copyright 2020 by the Institute of Physics.)

Ti:Sa light without locking to the seed, and up to 300 μJ of amplified light when the cavity was locked at a wavelength of 740.4 nm.

4.1.5 Z-Cavity Ti:Sapphire Laser

A Z-cavity layout Ti:Sa laser was used to provide the fundamental 740 nm BB light for the 246.8 nm transition. An example spectrum produced using this laser can be seen in Fig. 4.6, in red compared to the narrowband injection-seeded Ti:Sa, in blue and green. This cavity was designed and constructed at the University of Mainz [2, 9]. It has an operating range of 680–960 nm depending on the mirror set used, and a spectral linewidth of 3 GHz.

This Z-cavity laser was pumped by the LEE Nd:YAG laser and produced 240 μJ from 1.35 mJ of pump energy. The linear tripling setup was used for harmonic generation which was also shared by the injection-seeded Ti:Sa via the use of flip mirrors,

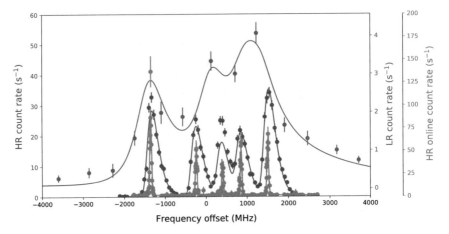

Fig. 4.6 Hyperfine spectra of atomic ^{115}In measured with the 246.8 nm transition under different conditions. (Red) Low-resolution scan using the Mainz Z-cavity Ti:Sa laser with indium ions created from the laser ablation ion source. (Blue) High-resolution scan with the injection-seeded Ti:Sa laser with indium ions created from the laser ablation ion source, before energy spread reduction. (Green) High-resolution scan with the injection-seeded Ti:Sa laser with indium ions from the ISOLDE ISCOOL cooler buncher

this allowed quick switching between BB and narrowband modes. See Fig. 4.2 for the layout.

4.1.6 M2 SolsTiS Tunable CW Ti:Sa Laser

The SolsTiS continuous-wave Ti:Sapphire laser by M-Squared was used as the seed of the injection-seeded laser for the 246.8 nm, 246.0 nm, 283.7 nm and 410 nm transitions. It is highly tunable with a range of 720–970 nm, linewidth of <50 kHz and output amplitude noise of less than <0.05%, making it ideal for producing seed light (Sect. 4.1.4).

Less than 2 mW was typically needed for seeding although it can output as much as 5 W with a higher power arrangement. The high-power light can be used to study many transitions with the CRIS technique [10, 11], by intensity modulation with a Pockels cell and polarisation selecting optics. The laser also has an integrated ECD-X doubling unit. The SolsTiS was pumped by 18 W of 532 nm light produced by a Sprout G-18W.

4.2 Higher Harmonic Generation Arrangements

Generation of light below 500 nm in these experiments was achieved with the use of non-linear crystals. For frequency doubling these are crystals with a $\chi^{(2)}$ nonlinearity [12].

Frequency tripling is not commonly achieved by a $\chi^{(3)}$ nonlinearity crystal material but instead by sum frequency generation. This is performed again in crystals with $\chi^{(2)}$ nonlinearity, in a cascaded process in which second harmonic light is generated and then mixed with the fundamental light in an additional crystal [13, 14]. BiBO (bismuth triborate BiB_3O_6) and BBO (β-barium borate, β-BaB_2O_4) crystals were used in the tripling and doubling setups for these experiments.

The intensity of the generated higher harmonic light grows as the square of the fundamental harmonic intensity, thus single-pass harmonic generation becomes feasibly efficient with the use of high-intensity pulsed laser light, 60 μJ (1.2 mJ/cm^2) was produced from 250 μJ (100 mJ/cm^2) of light from the Ti:Sa at 738 nm. Figure 4.3 shows the accessible wavelength ranges of Ti:Sa and dye lasers with higher-harmonic generation.

In the sum frequency generation arrangements used in these experiments type I phase matching was used, where the input harmonics require the same polarisation [15]. Two separate tripling setups were used with this in mind. A 'linear' tripling arrangement in which two non-linear crystals (BBO and then BiBO) share the same axis as the fundamental harmonic laser light, with an achromatic half-wave plate to rotate the polarisation of the fundamental and second harmonic light and a single lens to refocus them into the second BBO crystal for third harmonic generation. This has the advantage that the fundamental laser wavelength to be changed by a few nm and tripling can still be quickly recovered as the alignment through the arrangement is very simple. This arrangement was used to switch between the $5p\ ^2P_{3/2} \rightarrow 9s\ ^2S_{1/2}$ (246.8 nm) and $5p\ ^2P_{1/2} \rightarrow 8s\ ^2S_{1/2}$ (246.0 nm) transitions for these experiments, 230 μJ of fundamental light was able to generate 3 μJ of third harmonic light. The position of this linear arrangement can be seen in Fig. 4.2.

A separate 'split path' tripling arrangement was also used [16], in this case the path of the fundamental and second harmonic light is separated by a long-pass dichroic mirror. This allows the path, polarisation and focus of the two harmonic lights to be tuned independently to maximize overlap in the second BBO crystal, allowing higher efficiency third harmonic generation. For the $5p\ ^2P_{3/2} \rightarrow 5s5p^2\ ^4P_{5/2}$ (283.7 nm) transition it was only possible to generate 190 μJ of fundamental light due to the range of the injection-seeded Ti:Sa mirror coatings, but the efficiency of this split path arrangement still allowed 4 μJ to be delivered to the CRIS beamline.

4.3 Wavemeters and Reference Diode Laser

A WSU2 wavemeter by HighFinesse was used to measure the wavelength of the Ti:Sa laser light (including BB) in these experiments.

It has a specified accuracy of 2 MHz, if calibrated to reference light within 2 nm of the measured wavelength [17]. It is however subject to drift due to variations in temperature, and pressure in the ambient temperature of the laser lab. In order to counter this drifting a very stable reference laser at a fixed wavelength was used in combination with a four-channel fibre channel switchbox, which is able to rapidly sample the wavelength of several sources.

The reference diode laser used was a DLC DL PRO 780 by Toptica Photonics AG, this was used in combination with a compact saturated absorption spectroscopy [18] setup (CoSy FC-Cs+Rb+K), which provides Doppler-free hyperfine-absorption lines that the diode laser wavelength can be locked to. The quoted long term stability of the system is <1 MHz. In these experiments a hyperfine transition in rubidium was used [19].

The wavelength of the broadband pulsed dye laser scans was recorded using a HighFinesse WS6, quoted to have an absolute accuracy of 600 MHz in the range 420–1100 nm [17].

4.4 Transition Saturation

Saturation was determined by measuring the count rate at the resonance frequency at varying laser pulse energies and fitting with the saturation function [20]

$$I(P) = A_{jks} \frac{P/P_0}{1 + P/P_0} , \tag{4.1}$$

where P indicates the laser power and P_0 is the saturation power. A saturation curve is shown in Fig. 4.7 for the 5p $^2P_{3/2} \rightarrow$ 9s $^2S_{1/2}$ (246.8 nm) transition, where a laser fluence of 3.5(27) μJ/cm^2 was found to saturate the transition. The 5p $^2P_{3/2} \rightarrow$ 9s $^2S_{1/2}$ (246.8 nm) and 5p $^2P_{1/2} \rightarrow$ 8s $^2S_{1/2}$ (246.0 nm) transitions have strengths of $A_{jk} = 6.140 \times 10^6$ s^{-1} and $A_{jk} = 6.323 \times 10^6$ s^{-1} respectively [21]. Saturation was determined by using a half-wave plate to reduce the tripling efficiency using the 740.4 nm fundamental light, and then measuring the power of the 246.8 nm light at the final mirror into the CRIS beamline. The ion count rate was then measured at each power. Variations in atom beam intensity were taken into account by using the collisional background count rate, which is the primary source of uncertainty in this saturation measurement. Fluences much higher than 3.5(27) μJ/cm^2 were always used to ensure saturated conditions during these experiments, following this saturation measurement.

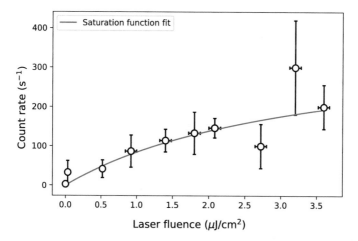

Fig. 4.7 Transition saturation curve for the 246.8 nm line, fit using Eq. 4.1 with orthogonal distance regression. A=380(190), $P_0 = 3.5(27)$ μJ/cm^2. The laser power was measured at the last mirror before reflection into the CRIS beamline

4.5   Timing Synchronization

The time synchronization of the laser pulses with the atom bunch in the interaction region of CRIS was achieved with the use of a Quantum Composers digital delay pulse generator (9520 Series). The timing resolution of 250 ps with less than 50 ps jitter which was sufficient for the precision required due to the lifetime of the atomic states studied in indium (3–50 ns).

A schematic example of the timing for the two-step 246.8 nm and 1064 nm ionization scheme is shown in Fig. 4.8, synchronising with the release of the indium ions from the cooler buncher ISCOOL. The final 1064 nm step created by the Nd:YAG laser head, a preceding delay of 510 μs was needed for the flashlamp relative to the Q-switch which released the light. The timing was such that the neutralised indium atoms were in the interaction region when the laser pulses passed through the beamline, with the non-resonant excitation step delayed by 20 ns to avoid power broadening or lineshape distortion of the transition peaks, with a reduction in laser ionization efficiency below experimental uncertainty when measured by count rate on resonance. A gate was then set on the MCP detection which accounted for the time of flight from the interaction region to the MCP detector, significantly reducing irrelevant background. For the ablation ion source experiments this was simply replaced with the Q-switch timing of the 532 nm Nd:YAG head which produced the indium ions.

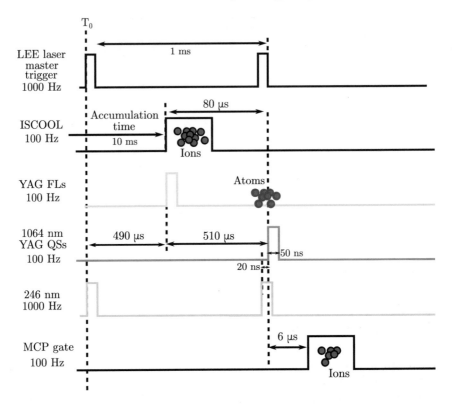

Fig. 4.8 Schematic of the timing synchronisation of the atoms and laser pulses in the interaction region for laser ionization and detection by an MCP detector. The relative timing of the resonant excitation step (246.8 nm/blue) and non-resonant ionization step (1064 nm/red) was also used to measure the lifetime of the excited state, see Table 5.9 for the results. QS: Q-Switch. FL: Flashlamp. Arrows indicate relative timing of pulses. Timing axis is not to scale

References

1. Kron T et al (2016) J Phys B: At Mol Opt Phys 49(18):185003. ISSN 0953-4075. http://stacks. iop.org/0953-4075/49/i=18/a=185003?key=crossref.b631c9d97f387b11727fe172cd8aba98
2. Rothe S et al (2011) J Phys: Conf Ser 312(5):052020. ISSN 1742-6596. http://stacks.iop.org/ 1742-6596/312/i=5/a=052020?key=crossref.11ec18510c5adb93eb3b4688e9627459
3. Zehentbauer FM et al (2014) Spectrochim Acta Part A: Mol Biomol Spectrosc 121:147. ISSN 1386-1425. https://www.sciencedirect.com/science/article/pii/S1386142513012195
4. Ready JF (1997) Industrial applications of lasers. Academic, Cambridge. https://books.google. ch/books?id=0BcKegyk87IC&dq=potassium+dideuterium+phosphate+switch+time& source=gbs_navlinks_s
5. Sonnenschein V et al (2017) Laser Phys 27(8):085701. ISSN 1054-660X. http://stacks.iop. org/1555-6611/27/i=8/a=085701?key=crossref.ae700ab1aa5d94924db40c8ed77febe7
6. Sonnenschein V (2015) PhD thesis, Department of Physics, University of Jyvaskyla
7. Kessler T et al (2008) Laser Phys 18(7):842. ISSN 1054-660X. http://link.springer.com/10. 1134/S1054660X08070074

8. Buczek C et al (1973) Proc IEEE 61(10):1411. ISSN 0018-9219. http://ieeexplore.ieee.org/document/1451224/
9. Mattolat C et al (2009) In: AIP Conference Proceedings, vol 1104, pp 114–119. AIP. ISSN 0094-243X. https://doi.org/10.1063/1.3115586
10. de Groote RP et al (2015) Phys Rev Lett 115(13):132501. ISSN 1079-7114. https://doi.org/10.1103/PhysRevLett.115.132501, https://doi.org/10.1103/PhysRevLett.115.132501
11. Voss A et al (2013) Phys Rev Lett 111(12):122501. ISSN 0031-9007. https://doi.org/10.1103/PhysRevLett.111.122501
12. Franken PA et al (1961) Phys Rev Lett 7(4):118. ISSN 00319007
13. Bass M et al (1962) Phys Rev Lett 8(1):18. ISSN 0031-9007. https://link.aps.org/doi/10.1103/PhysRevLett.8.18
14. New GHC et al (1967) Phys Rev Lett 19(10):556. ISSN 0031-9007. https://link.aps.org/doi/10.1103/PhysRevLett.19.556
15. Eckardt R et al (1984) IEEE J Quantum Electron 20(10):1178. ISSN 0018-9197. http://ieeexplore.ieee.org/document/1072294/
16. Wilkins S (2017) PhD, University of Manchester
17. HighFinesse (2019) Wavelength meter specifications. Technical report, HighFinesse. https://www.toptica.com/fileadmin/Editors_English/11_brochures_datasheets/03_HighFinesse/HighFinesse_Wavemeter.pdf
18. Hall JL et al (1976) Phys Rev Lett 37(20):1339. ISSN 0031-9007. https://link.aps.org/doi/10.1103/PhysRevLett.37.1339
19. Steck DA (2001) Rubidium 87 D line data. Technical report, Los Alamos National Laboratory, Los Alamos, NM 87545. http://steck.us/alkalidata%5Cn, https://doi.org/10.1029/JB075i002p00463
20. Siegman AE (1986) Lasers. University Science Books, Mill Valley, California. http://www.worldcat.org/title/lasers/oclc/14525287
21. Kramida NATA, Ralchenko Y, Reader J (2014) NIST atomic spectra database. http://www.nist.gov/pml/data/asd.cfm

Chapter 5
Laser Ablation Ion Source Studies of Indium

5.1 Introduction

Laser spectroscopy experiments at radioactive beam facilities make measurements on elements across the periodic table [1–5]. This makes choosing a single ion source for 'offline' testing with stable isotopes difficult as most sources are only able to produce ions with specific properties e.g. alkali metals or volatile elements. Furthermore the efficiency and sensitivity of the CRIS technique [6] depends greatly on a bunched time structure of the ion beam, which is usually provided by a cooler buncher. Or alternatively by chopping [7] of a continuous ion beam, however this comes at a huge cost in duty cycle loss in terms of the measurement of the total number of produced ions.

This section will focus on work also partly covered in the paper [8], which is dedicated to the characterisation of a laser ablation ion source, which was found to be ideal for laser spectroscopy experiments and enabled high-precision measurements of atomic physics parameters when used in tandem with the CRIS technique. The ablation process intrinsically provides the pulsed time structure, high peak ion intensity and synchronization with the subsequent lasers. This provides the ideal conditions for high spatial and temporal laser-atom overlap to perform efficient CRIS. The ablation ion source is able to produce ions of almost any element and in many initial charge states [9], provided that element can be contained in a solid matrix to be ablated [10, 11]. Targets of copper, indium, tin and aluminium were mounted at the same time which allowed production of ions of each element with only a few minutes needed to switched between targets.

Measurements began on indium as a way to characterise the feasibility of an ablation ion source for offline tests at the CRIS experiment and develop an optimal ionization scheme for the following measurements on neutron-rich indium isotopes. An extraction potential configuration was eventually found allowing measurements of hyperfine spectra with linewidths of 160(30) MHz. It was determined that the

© The Editor(s) (if applicable) and The Author(s), under exclusive license
to Springer Nature Switzerland AG 2020
A. R. Vernon, *Collinear Resonance Ionization Spectroscopy
of Neutron-Rich Indium Isotopes*, Springer Theses,
https://doi.org/10.1007/978-3-030-54189-7_5

magnetic dipole hyperfine structure constants A_{hf} of the some the upper states being measured were of sufficient precision to benchmark atomic coupled-cluster calculations [12], previous results of which highlight atom indium as a candidate system for electron electric dipole moments [13] (eEDM), due to a large EDM enhancement factor in indium. These states were uniquely accessible to the CRIS experiment due to the population of meta-stable states in charge exchange (see Sect. 2.4) and the ability to produce high-resolution deep-UV light [3]. Further improvements which ultimately lead to hyperfine spectra with linewidths of 40(10) MHz and improvement of the 'CCSD(T)' calculations (see Sect. 2.3.2) in collaboration with an atomic theorist are detailed in the sections below.

5.1.1 Ablation Ion Source Arrangement

A versatile ion optics arrangement was constructed to accommodate multiple types of ion source to cover the range of elements needed for 'offline' testing using stable isotopes with the CRIS setup. This arrangement is shown schematically in Fig. 5.1, and was tested to be capable of refocusing and extracting ions, from a surface [14, 15] ion source or Colutron-type [16, 17] plasma ion source, with the sources mounted in front of the first extraction plate, labelled as 'Ext1' in Fig. 5.1.

The surface ion source is simply a heated capillary tube. These type of surface ion sources have a typical energy spread of less than 1 eV [18]. The tantalum surface ion source used at the CRIS experiment can ionize elements with ionization potentials below 5 eV (the work function of tantalum is 4.5 eV) [19], allowing the ionization of alkali metals. The Colutron plasma ion source has an energy spread of <0.1 eV [20] and is capable of delivering μA beams. However it requires the element to be gaseous, and its electron emitting filament has a limited operational lifetime of around a day.

Between two of the extractor plates a retractable target ladder was installed at 45° to entrance and exit laser windows. As shown in Fig. 5.1a). This ladder held the indium target material for laser ablation. At the other positions of the ladder different materials were held, including aluminium, copper and tin, a time-of-flight spectrum of these elements extracted from the ion source is shown in Fig. 5.2.

The ion optics arrangement also incorporates an Einzel lens [21] to focus into the extractor 'Ext 4', held at ground relative to the preceding components of the ion source. These preceding components were held at up to 30 kV at one side of a PEEK insulator beam pipe. The whole arrangement was contained inside of a high-voltage cage. Following extraction to beam energies of up to 30 keV, electrostatic deflectors were mounted to make adjustments to the trajectory of the exiting beam into a large Einzel lens. This lens was held at around -13 kV and provided focussing into 90° electrostatic bending plates mounted downstream, which guided the ions into the CRIS beamline. The layout of the source in relation to the interaction region is shown in Fig. 4.2. The simulated trajectories of the ions extracted from the ablation ion source are shown in Fig. 5.12 and discussed in Sect. 5.4.1. The electronic setup

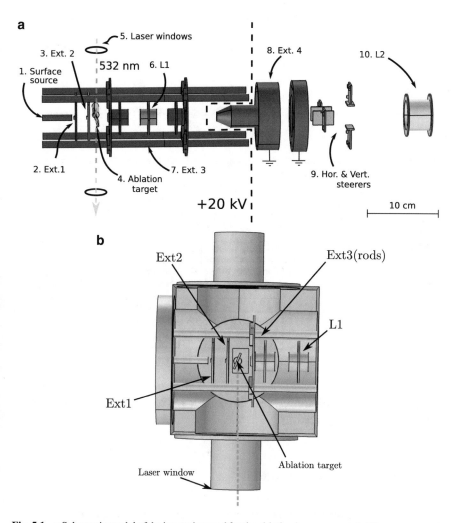

Fig. 5.1 a Schematic model of the ion optics used for the ablation ion source work. The arrangement is also used for surface and plasma sources. A nominal voltage of +20 kV was was used, but up to +30 kV was also possible with the ablation setup. **b** Corresponding rendering of the chamber vacuum and ion optics near the ablation target extraction point. Laser windows allow 532-nm light to be focused onto the ablation target

used to control the potentials of the ion source is shown in Fig. 5.3. The technical details are described in [22].

Another potential application of this ion optic arrangement is as an atomic beam unit [23], for Doppler-free laser spectroscopy. This is made possible when the polarity of the extractor plates is inverted, repulsing the positive surface ions and allowing

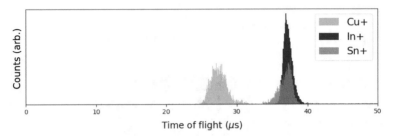

Fig. 5.2 Comparison of the time of flights of the different elements mounted on the ablation ion source ladder. A binning of 100 ns was used

Fig. 5.3 Diagram of the electronics for the isolated HV control setup. See proceedings [22] for details of the control

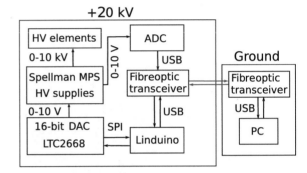

only the neutral fraction from the surface source to pass perpendicular to the laser windows for laser spectroscopy.[1]

5.2 Experimental Method

The ablation process is known to be complex and contains many regimes of ionization [24–26]. The resultant beam was expected to have a large energy spread, spreads of 0.1–1 keV are typically observed in the ablation process [27].

An aim of this work was to investigate the feasibility of using this type of ion source for high-precision measurements, this requires energy spreads of <2 eV to be within the linewidth of the injection-seeded Ti:Sa laser. The characteristics of the extracted ion beam in an ablation ion source has a strong dependence on the surrounding electric field, which interacts with the ablation plasma plume. This was explored by varying the voltages applied to the elements surrounding the ablation target, and measuring the alteration to the time of flight (ToF) and energy spread of the ions, seen in hyperfine spectra.

---

[1] Reprinted from Vernon, A. R. et al. Optimising the Collinear Resonance Ionisation Spectroscopy (CRIS) experiment at CERN-ISOLDE. NIM B. Copyright (2019) with permission from Elsevier.

5.2.1 Detection Configuration

The ablation ion source and extraction optics were held at 20 kV relative to the CRIS beamline. A pair of 90° bending plates were used to deflect the ions into the section of the beamline before a 34° bend to overlap with the lasers collinearly. The layout of the beamline can be seen in Fig. 3.6.

The ions were then neutralised with a sodium-filled charge-exchange cell at 300(10) °C. An iris after the charge-exchange cell and another after the interaction region were used to align the atom-laser overlap.

A micro-channel plate [28] (MCP) detector was placed in the path of the ions deflected by 20° after ionization for direct detection. The MCP output signals were recorded by a TimeTagger4-2G time-to-digital data-acquisition card with 500 ps time resolution on the ion time of flight (ToF). The ablation laser Q-switch was used as the timing start trigger. A 4 μs gate around the bunch was used in analysis to suppress irrelevant background counts from the MCP signals.

5.2.2 Laser Configuration

The 246.0-nm ($5s^2 5p\ ^2P_{1/2} \to 5s^2\ 8s\ ^2S_{1/2}$), 246.8-nm ($5s^2\ 5p\ ^2P_{3/2} \to 5s^2\ 9s\ ^2S_{1/2}$), 283-nm ($5s^2\ 5p\ ^2P_{3/2} \to 5s^2\ 5s5p^2\ ^4P_{5/2}$) and 410-nm ($5s^2\ 5p\ ^2P_{1/2} \to 5s^2\ 6s\ ^2S_{1/2}$) transitions were used to measure the hyperfine structures of the ^{113,115}In isotopes produced using the ablation ion source. The details of the spectroscopy lasers used and the laser lab arrangement for these measurements is detailed in Chap. 4. The full schemes for ionization following the excitation steps is shown in Fig. 4.1.

The laser beam diameter in the interaction region was ∼8 mm, giving a fluence of 120 mJ/cm^2 for the 1064-nm light and 60 mJ/cm^2 for the 532-nm light. The second Litron Nd:YAG laser head was used to produce 532-nm light for the laser ablation. It had focussed spot size of <1 mm set on the ablation target, corresponding to a fluence of >0.5 J/cm^2. The flashlamp of this laser head was used as the start trigger for the time-of-flight measurement, whilst the other lasers were offset in time from this trigger to meet the atom bunch in the interaction region. The time synchronization was achieved using the Quantum Composers digital delay pulse generator (Sect. 4.5).

Figure 5.4 shows the ion beam current produced from the ablation source as a function of time from initiating the 532-nm ablation laser. This was measured at a Faraday cup directly after the 90° bend following the ion source (FC90). Fluctuations on a level of 30% of the ion beam intensity were observed on a 10 s time scale for the first 5 min, after 30 min the fluctuations were on the level of 7%. A rolling average of 1 min shows no fluctuations over a slow plateauing in beam current. The transmission efficiency from the 90° bend to a Faraday cup (FC3) at the end of the interaction region was optimised to and maintained at a value of 10%. Laser frequency scans over the hyperfine structure were measured at least an hour after switching on the

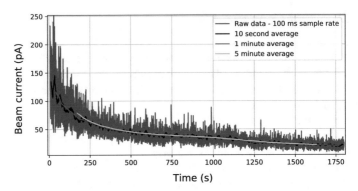

Fig. 5.4 Ion beam current measure over 30 min on a Faraday cup at the 90° bend of the ion source. The ablation laser was started at $T = 0$ s. Rolling averages are shown for 10 s, 1 min and 5 min

ablation laser. Rotating targets are routinely used in ablation ion sources to improve beam-current stability and intensity [29, 30]. This will be implemented in the near future for this ablation setup.

5.3   Analysis Procedure

5.3.1   Systematic Errors

In order to quantify possible sources of systematic error, the variation in extracted hfs constants was explored using different experimental conditions and data analysis procedures, following the same approach as described in Sect. 6.3.2 for the neutron-rich indium isotopes. With the exception that the Doppler-shift uncertainty was based on the ion source high-voltage supply instead of the platform voltage of ISCOOL. The same fitting procedure as for the neutron-rich indium isotopes, described in Sect. 6.1 was also used for this work. The laser-frequency binning procedure required modification and is described in Sect. 5.3.2 below.

The quantified uncertainties are displayed in Table 5.1, where the the hyperfine structure parameters uncertainties were evaluated separately for each atomic state for ^{115}In and ^{113}In. Due to the order of magnitude difference in natural abundance between the isotopes, the hyperfine spectra measured for one mass may be more susceptible to a source of error than the other, therefore the uncertainties were quantified for each mass separately. For example slight fluctuations in collisional background or laser intensity would have a more significant effect on the position or shape of the smaller ^{113}In hyperfine structure peaks, as the ^{113}In spectra took longer to gain the same level statistics as ^{115}In. The distribution of systematic error varied for each transition used, due to a combination of varying scan quality, magnitude of hyperfine

Table 5.1 Principle sources of systematic error identified in extracting hfs constants and isotope shifts from the measured hfs spectra

	A_{hf} (MHz)						B_{hf} (MHz)		$\delta\nu^{115,113}$ (MHz)		
	5p $^2P_{1/2}$	5p $^2P_{3/2}$	6s $^2S_{1/2}$	8s $^2S_{1/2}$	9s $^2S_{1/2}$	5s5p^2 $^4P_{5/2}$	5p $^2P_{3/2}$	5s5p^2 $^4P_{5/2}$	246.0 nm	246.8 nm	410.2 nm
^{113}In											
Ion Energy	0.10	0.04	0.16	0.06	0.04	0.82	0.09	2			
Wavemeter cal.	0.74	0.19	0.07	0.15	0.24	3.1	0.82	4			
Binning	0.19	0.09	0.74	0.17	0.29	1.2	1.10	3			
Total error	0.77	0.21	0.76	0.23	0.38	3.42	1.37	5			
^{115}In											
Ion Energy	0.02	0.004	0.12	0.006	0.016	0.3	0.01	0.20	3.7	4.1	4.8
Wavemeter cal.	0.53	0.26	0.37	0.19	0.14	2.3	0.44	2.6	2.1	4	1.5
Data analysis	0.04	0.016	0.031	0.021	0.044	0.4	0.15	1.8	0.46	0.13	0.53
Δt									1.6	2.7	1.3
Total error	0.53	0.26	0.39	0.19	0.15	2.3	0.46	3.2	4.3	5.7	5.1

constants and number of peaks in the hyperfine spectra. The total systematic error for ^{115}In was found to be less than for ^{113}In for all but one state.

There was an additional source of error on the isotope shift due to the uncertainty on the time-of-flight difference between ^{115}In and ^{113}In, which was needed to perform time-of-flight based corrections according to the procedure described in the following Sect. 5.3.2. The time-of-flight difference between ^{115}In and ^{113}In is denoted as Δt in Table 5.1, and was determined by refitting of the data with the range of Δt values within the uncertainty range of the ToF.

5.3.2   Time-of-Flight Lineshape Correction

For most spectra with laser limited linewidth resolution in this work, such as those shown in Figs. 4.4 or 6.1, the count rate of ions incident upon the MCP detector was correlated with the frequency of the scanning laser by timestamp and then the counts were binned with respect to laser frequency. The binning procedure in that case is explained in Sect. 6.1.

In this ablation ion source work the ToF of the laser-ablated ions was additionally recorded, with 500-ps time resolution. This revealed a relationship between the ToF and the Doppler shift of the detected ions, as can be seen in Fig. 5.5.

The shape of this relationship was found to be dependent on the extraction field around the ablation target. In the case presented in Fig. 5.5, the linewidth would be limited to 300–400 MHz without the time-of-flight information. But as can be seen in orange, a time-of-flight slice is able to significantly improve the resolution and remove the peak asymmetry but at the cost of reduced statistics, for spectra constructed from a single time-of-flight selection.

Fig. 5.5 Combined surface plot and histograms for laser frequency versus ToF. ^{115}In measured with the $5p\,^2P_{3/2} \rightarrow 9s\,^2S_{1/2}$ (246.8-nm) transition for an electric field gradient of 5 V/mm around the ablation plume. ToF slices (pale orange) can be used to create a Doppler-corrected final hfs spectrum (blue). Without the ToF data the hfs resolution has a significantly larger linewidth (orange). © 2020 American Physical Society

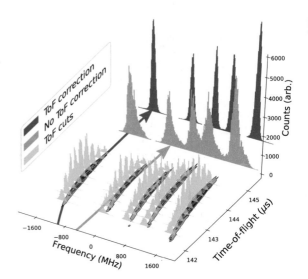

It was found to be possible to use these time-of-flight slices to fit the relationship for a given field configuration in order to better utilise the data of a scan and significantly improve the resolution of a spectrum.[2]

The correction was performed by taking 100-ns slices of this data set and fitting the centroids of the hfs spectrum. The centroid values were then used to construct a spline in order to interpolate the frequency correction needed for the ToF of each ion count. To first order the hfs constants are independent of the atom beam energy at which they were measured, therefore each slice should have the same hfs constant. A linewidth of 40 MHz was achieved with this process, limited by a compromise between the linewidth of the Ti:Sa laser used (34 MHz [31]) and the number of ion counts split into each slice to bin and fit a spectrum.

Figure 5.8 demonstrates that the fitted hyperfine structure constants from the individual slices are within statistical error of eachother and that the average value of the slices usually agrees well with the value of the corrected and recombined spectrum for each scan. The final values for the $5p\,^2P_{3/2}$ state from this study with the ablation ion source are $A_{hf} = 241.98(38)$ MHz and $B_{hf} = 450.0(1.5)$ MHz, which agree with the literature values of $A_{hf} = 242.1647(3)$ MHz and $B_{hf} = 449.545(3)$ MHz. The average values are indicated in grey in Fig. 5.8. In a few scans the average from the slices does not agree with the final value from this work but the combined spectrum does, this is likely due to an additional error introduced by fitting the hyperfine spectra to the slices with reduced statistics, which is remedied after the correction.

Although to first order the hyperfine structure constants are independent of the ToF slice, ^{113}In will have a different ToF to ^{115}In and so a given ToF slice will

---

[2] Reprinted with permission from Garcia Ruiz, R. F., Vernon, A. R., et al. High-Precision Multiphoton Ionization of Accelerated Laser-Ablated Species. Phys. Rev. X 8, 041005 (2018). Copyright 2020 by the American Physical Society.

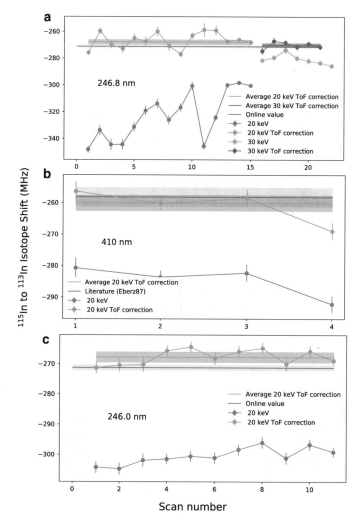

Fig. 5.6 Isotope shift values of individual scans for the **a** $5p\ ^2P_{3/2} \rightarrow 9s\ ^2S_{1/2}$ (246.8-nm), **b** $5p\ ^2P_{1/2} \rightarrow 6s\ ^2S_{1/2}$ (410-nm), literature from [32] and **c** $5p\ ^2P_{1/2} \rightarrow 8s\ ^2S_{1/2}$ (246.0-nm) lines showing the average, before (blue) and after (orange) including the ^{115}In, ^{113}In ToF difference in the correction procedure. © 2020 American Physical Society

not correspond to the same Doppler-shifted atomic ensemble for both masses, this would invalidate the extraction of a precise isotope shift. To correct for this a separate additional spline shifted by the difference in the ToF between ^{113}In and ^{115}In, $\Delta t\,(^{115}$In $-^{113}$In), was used for the ToF correction of the ^{113}In spectra. Therefore for each count event with a time of flight recorded while in the scan region of ^{113}In, C_t, the frequency correction applied for the combined binning is (Fig. 5.6)

$$\nu_{113\text{In}} = \nu_{115\text{In}}(C_t - \Delta t(^{115}\text{In} - ^{113}\text{In})) \,, \qquad (5.1)$$

where $\nu_{113\text{In}}$ is the assigned frequency of the count event and

$$\nu_{115\text{In}}(C_t - \Delta t(^{115}\text{In}\text{-}^{113}\text{In})) \,,$$

is the correction used from ^{115}In. This correction was permitted as the ToF was cleanly separable by laser frequency due to the kinematic and isotope shift between ^{113}In and ^{115}In. Figure 5.7 illustrates the laser frequency selection made to determine the ToF difference between ^{113}In and ^{115}In at the ionization point. This result is interesting in its own right, as the mass resolving power in the ToF spectrum is dominated by the laser selectivity associated with RIS.

The selectivity, S, is given by

$$S = \left(\frac{\delta\omega}{\Gamma}\right)^2 \,, \qquad (5.2)$$

where $\delta\omega$ is the frequency separation between the two isotopes and Γ is the resolution of the hyperfine structure transitions. At 20 keV, a selectivity of 2×10^4 was reached during these experiments. The selectivity at 20 keV is $\sim$2000 times higher than at rest (as was illustrated in Fig. 3.7).

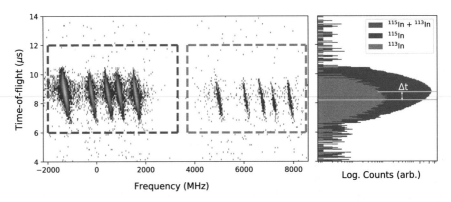

Fig. 5.7 Frequency selection for isotope time of flight. Ion count rate is indicated by the intensity of the surface plot as a function of the laser frequency (x axis) and ion ToF (y axis). This histogram on the right displays the frequency gated ToF distributions for ^{115}In (blue) and ^{113}In (green). $\Delta t_{115,113}$ = 0.36 μs for 20 keV. © 2020 American Physical Society

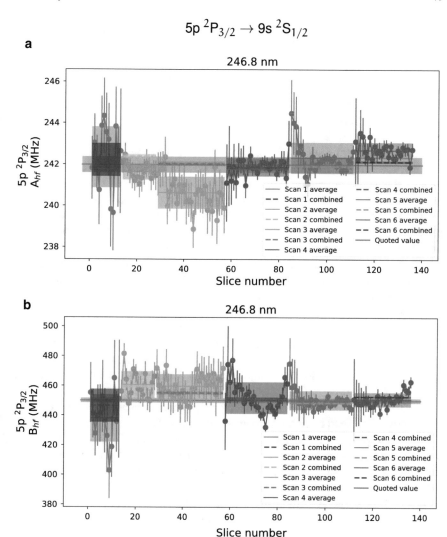

Fig. 5.8 Comparison of fitted hyperfine parameters from taking the average of several ToF slices of around 0.5 µs versus the fitted hyperfine parameter after the ToF correction. The quoted value was the average many more scans than is shown. All scans were taken at 20 keV

Figure 5.6 shows the effect of the shifted-ToF-frequency curve for ^{113}In for the lines measured in this work. The correction worked as expected and bought the extracted isotope shift values into agreement with literature values [32] and the online values later determined from the neutron-rich isotope measurements (Sect. 6.3) (Fig. 5.8).

5.4 Results

5.4.1 Ablation Ion Source Extraction Field

The influence of the electric field gradient surrounding the ablation target on the energy spread of the produced ion beam was explored by applying the voltage configurations in Table 5.2.

The ablation laser fluence of >0.5 J/cm^2 should correspond to an energy spread of 7 eV according to the compilation by [27]. However due to the low number of ions created at this energy there was no previously experimental data available, this goes some way in explaining the lower energy spread observed in this work.

A low transmission efficiency and large energy spread for any potential applied to elements 'Ext1' or 'Ext2' behind the ablation target (Fig. 5.1a, b) was initially observed. An example of newa spectrum taken in this mode is given in Fig. 5.9, where the linewidth became 375(25) MHz. This gave a first hint that understanding the systematics of the target extraction field would be important for obtaining high-precision results from the ablation setup.

The high- and low-field configurations labelled 'cfg 1' and 'cfg 2' respectively, were found to be two exemplary cases for understanding the process. The high extraction field case was found to have a gradient of 5 V/mm by simulation and the low field case was found to have a gradient of 0.01 V/mm. The simulations were performed in COMSOL Multiphysics [34]. Figure 5.10 shows the results as surface plots. The time-of-flight distributions for the two configurations are shown in Figs. 5.10.A2 and 5.10.B2, showing that the distribution arrives earlier for the higher extraction voltage and has a tail towards slower velocities.

The corresponding hyperfine spectra are shown in Figs. 5.10.A3 and 5.10.B3 where a significant asymmetry was observed for the high-field configuration with linewidths that ranged from 80 to 120 MHz, this was attributed to the difference in velocity distribution of ions generated in these two configurations from the extraction field geometries. The velocity distribution generated in the ablation process can be described by a modified Maxwell–Boltzmann distribution which includes thermal

Table 5.2 Combination of voltages applied to the ablation ion source optics to explore field effects. Refer to Fig. 5.1 for a schematic of the ion optics. Ext1 and Ext2 were grounded. †—the absolute beam energy was determined to be $T_B = 19989(1)$ eV from the Doppler shift of the 5p ^{2}P$_{3/2}$ → 9s ^{2}S$_{1/2}$ (246.8-nm) transition frequency

cfg	Ext 3/rods (V)	L1 (V)	L2 (kV)	Field gradient (V/mm)	T_B† (keV)
1	−90	−600	−13.61	5	20
2	0	−250	−12.68	0.01	20
3	−160	−1250	−12.69	8	20
4	−50	−480	−12.69	3	20

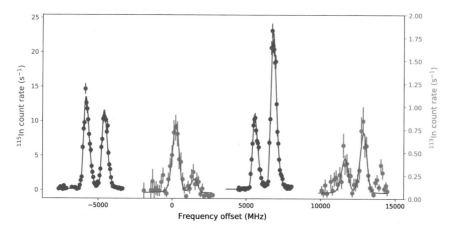

Fig. 5.9 A hyperfine spectrum of ^{115}In (blue) and ^{113}In (green) taken with the 5p ^{2}P$_{3/2}$ → 9s ^{2}S$_{1/2}$ (246.8-nm) line. Fits using an asymmetric Voigt profile are shown by the solid lines. Large asymmetric linewidths of 375(25) MHz were observed due to a leakage potential on the elements surrounding the ablation target. These spectra were fitted with an asymmetric pseudo Voigt [33]

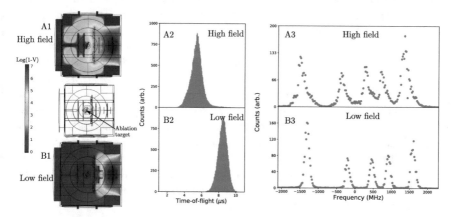

Fig. 5.10 Influence of the extraction field around the ablation target on the ion ToF and energy spread. Row A(B) shows the high(low) field configuration of 5 V/mm(0.02 V/mm). The field gradient simulation is shown in part A1(B1), voltages are with respect to the local ground of the source. The resulting ToF for the field is shown in part A2(B2) and the resulting hyperfine spectrum in part A3(B3). © 2020 American Physical Society

effects and a higher energy distribution due to the laser-generated plasma effects [25, 27, 35] (Fig. 5.11).

Ion transport simulations [34] for the two field geometries discussed are shown in Fig. 5.12. A hemispherical ion source was used to approximate the ions created by ablation with an energy spread of 10 eV, the minimum energy spread that could be detected with the 34 MHz laser linewidth after acceleration to 20 keV. Although this simulation neglected much complexity of the ablation process two trends were

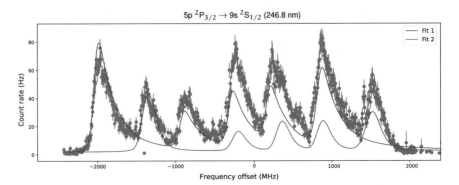

Fig. 5.11 Asymmetry observed in hyperfine spectrum of ^{115}In using an extraction field gradient of 5 V/mm and a beam energy of $T_B = 20$ keV. Two overlapping hyperfine spectra components appear when ToF is not taken into account at high field gradients. Fitted using two asymmetric pseudo-Voigt profiles

reproduced; around 4 μs difference in the ion ToFs and transmission of a narrower initial velocity distribution by the low-gradient configuration. These features are visible in Fig. 5.10.B and in the ion scatter in Fig. 5.12.B.

Besides understanding how to reduce the energy spread and therefore linewidth of spectra taken with the ablation ion source, the data gained with each field configuration combined with laser spectroscopy gives a new perspective on relationship between the velocities and energy spread from the laser plasma. Figure 5.13 illustrates this dependence. It is intuitive that a higher extraction field corresponded to a shorter ToF, as the extraction time inside the source before acceleration to 20 keV is a significant contribution to the final ToF and depends on the field gradient. Additionally, the curvature towards lower energy spread and the more uniform velocity at the slower tail of the ToF distribution gives a direct measurement of the bimodal slow thermal and prompt ablation plasma velocity distributions often used to model the laser ablation process [25].

5.4.2   Atomic Parameters

5.4.2.1   Extraction of Parameters

Spectra for the near-zero extraction field and after ToF corrections are shown in Fig. 5.15 for the $5p\,^2P_{1/2} \rightarrow 8s\,^2S_{1/2}$ (246.0-nm), $5p\,^2P_{3/2} \rightarrow 9s\,^2S_{1/2}$ (246.8-nm), $5p\,^2P_{3/2} \rightarrow 5s5p^2\,^4P_{5/2}$ (283.7-nm) and $5p\,^2P_{1/2} \rightarrow 6s\,^2S_{1/2}$ (410-nm) transitions. A full scan of the hyperfine structure for the $5p\,^2P_{1/2} \rightarrow 8s\,^2S_{1/2}$ (246.0-nm) and $5p\,^2P_{3/2} \rightarrow 9s\,^2S_{1/2}$ (246.8-nm) transitions took around 5 min at a rate of 15 MHz/s. The larger $5p\,^2P_{1/2} \rightarrow 6s\,^2S_{1/2}$ (410-nm) ($\sim$30 GHz) and $5p\,^2P_{3/2} \rightarrow 5s5p^2\,^4P_{5/2}$ (283.7-nm) ($\sim$ 100 GHz) structures took 20 min and 1 h respectively to scan.

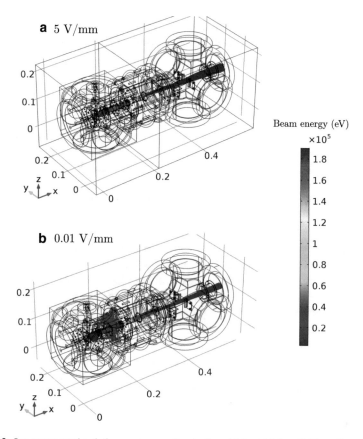

Fig. 5.12 Ion transport simulations corresponding to the **a** high and **b** low field gradient configurations of Fig. 5.10. Spacial units in meters

Fig. 5.13 Centroids of hyperfine spectra sliced and fitted in ToF in four extraction voltage regimes. c.f. Table 5.2. Width of bars represents the FWHM of the hyperfine spectrum slice. All of the spectra used were taken at a beam energy of 20 keV. © 2020 American Physical Society

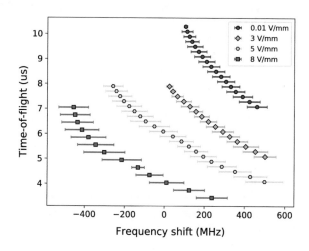

From Faraday cup readings, a beam current of 300(100) fA was measured directly after the charge-exchange cell, with a transmission efficiency of 50% through the cell, therefore an estimated $10^5 - 10^6$ ion/s were entering the interaction region. A mean count rate of 500/s was observed on resonance in an example scan, this gives an efficiency of >0.05% for the whole detection process (Sect. 3.2). Which was a sufficient magnitude for detection in this work. With a maximum expected efficiency of ~60%, limited by the population following charge exchange (see Fig. 5.20), the remaining loss in efficiency could have been due to poor atom-laser overlap, low resonance ionization efficiency, ion transmission to the MCP detector, or a combination of all of these. The signal-to-background rates for ^{115}In and ^{113}In were found to be 10^4 and 10^3, with the difference due to the lower relative natural abundance of ^{113}In.

An example fit to a hyperfine structure spectrum with residuals is shown in Fig. 5.14a) before including the ToF correction a linewidth of 110(30) MHz was obtained using a Voigt profile. This is compared to after including the including the ToF correction in Fig. 5.14b) where a linewidth of 68(1) MHz was found. This spectrum was measured using the low field voltage configuration (0.02 V/mm). χ_r^2 values of 1-2 were usually obtained for each fit after including the correction. Table 5.3 displays the FWHM and asymmetry parameter (0 corresponding to no asymmetry and 1 to an infinite tail [33]) for asymmetric pseudo-Voigt line profiles [33] fitted to ^{115}In hyperfine spectra with varying extraction field gradients. As previously highlighted, the FWHM and asymmetry increases with increasing field gradient. At the field gradient of 3 V/mm and above the asymmetry becomes too large for the asymmetric pseudo-Voigt profile as additional hyperfine spectra appear, corresponding to the other time-of-flight components which are extracted from the ablation source. See Fig. 5.11. The remaining Gaussian and Lorentzian contributions to the FWHM of the Voigt profiles of the spectra in Figs. 5.14 and 5.11 after the ToF correction, along with the parameters from other field gradient configurations are also shown in Table 5.3.

A clear correlation between the electric field gradient and the Gaussian contribution to the linewidth was observed. This can be attributed to a residual energy spread from the field gradient. Each field gradient required adjustment of the subsequent ion optics to regain a nominal 300(100) fA beam current used for these measurements, measured at the end of the interaction region. Therefore the beam intensity remained the same while the energy spread of the bunch increased, in the analysis 100 ns slices were used for all configurations and therefore the remaining Gaussian linewidth increased, this is reflected in the final correct values in Table 5.3. This can be seen in the FWHM of Figure 5.13. Hence with increased beam current, finer time slices could be used and reduced Gaussian contributions obtained. However from the linewidth obtained, an upper limit on the intrinsic energy spread of the bunch in the high-field condition of 8 V/mm can be given a value of <13 eV. Table 5.4 gives the relative FWHM contributions as a function of beam energy, T_B. The Gaussian contribution was observed to reduce with increasing T_B, as the energy spread of the bunch is reduced. From this measurement. The intrinsic energy spread of the bunch created by the ablation laser under the low field condition can be given a value of <3.5 eV from these measurements.

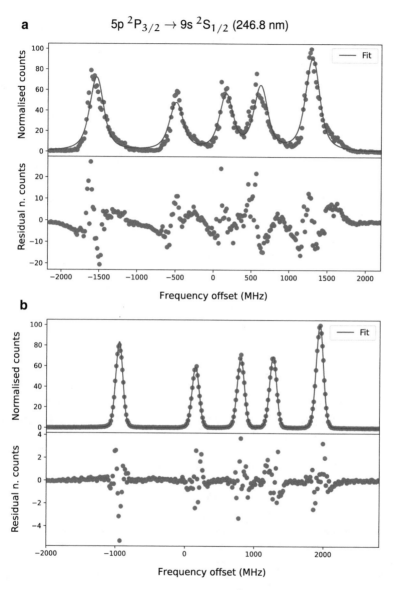

Fig. 5.14 An example fitted hyperfine spectrum with the $5p\,^2P_{3/2} \to 9s\,^2S_{1/2}$ (246.8-nm) transition, showing residuals. Part **a** shows the spectrum without correcting for ToF, part (linewidth of 150(30) MHz) **b** shows the spectrum after correction for ToF (linewidth of 68(1) MHz), as described in Sect. 5.3.2. The spectrum was measured with the 0.02 V/mm extraction field configuration. The total count rate was normalised to 100. A Voigt line profile was used

Table 5.3 Lineshape parameters of the asymmetric pseudo-Voigt profile used to fit spectra obtained under varying extraction field gradients (Field grad.), then the remaining Lorentzian (FWHM L.) and Gaussian (FWHM G.) linewidth contributions to a Voigt fit after the ToF correction. Refer to Table 5.2 for the voltage configurations

		ToF uncorrected		ToF corrected	
Field gradient (V/mm)	cfg #	FWHM Asym. (MHz)	Asym P. (MHz)	FHWM Gaus. (MHz)	FHWM Loren. (MHz)
0.01	2	162(35)	0.63(3)	65.6(6)	13.8(4)
3	4	279(62)	0.61(10)	78.8(9)	16.9(6)
5	1	>280*	>0.7*	121(1)	16.5(7)
8	3	>280*	>0.7*	146(1)	26.6(9)

*—Fitting with a single asymmetric pseudo-Voigt profile no longer works at these high field gradients, see Fig. 5.11

Table 5.4 The remaining Lorentzian (FWHM L.) and Gaussian (FWHM G.) linewidth contribution to a Voigt lineshape as a function of beam energy T_B

T_B (keV)	FHWM Gaus. (MHz)	FHWM Loren. (MHz)
10	59(6)	19(5)
20	53.7(5)	14.7(4)
30	47(1)	13.2(7)

In both Tables 5.3 and 5.4 the Lorentzian component is smaller than the laser linewidth ($\sim$35 MHz). In both cases this is likely due to a non-uniform energy spread with ToF, creating an effectively larger Gaussian component when the spectra are corrected, which may explain the remaining residuals to the Voigt fit seen in Fig. 5.14. This could be corrected by using smaller time slices to remove the varying Gaussian contribution, this would also require a larger beam current. Alternatively the Δt of the time slices could be chosen to be non-uniform in order to select parts of the bunch with a uniform energy spread (Fig. 5.15).

Figure 5.16 compares the final average values of the extracted hyperfine structure constants for the different field configurations before and after the ToF correction. The four varied approaches were found to agree with each other and also the final quoted values from this work. The approach with the least uncertainty was for the zero-field configuration after ToF correction, which could be expected from from the reduced asymmetry and linewidth of the uncorrected spectra.

The final quoted values for the hyperfine structure constants of ^{115}In and ^{113}In are presented in Table 5.7. These values agree well with previous measurements for the $^2P_{1/2,3/2}$ states which were available. For the $^4P_{5/2}$ state two measurements of A_{hf} have been made [36, 37] which are 2σ apart, the measurement from this work agrees closest with the laser spectroscopy measurement by [37] and improves on its

Fig. 5.15 Measured
hyperfine structures of ^{115}In
(blue) and ^{113}In (green) for
the **a** 5p ^{2}P$_{1/2}$ → 8s ^{2}S$_{1/2}$
(246.0-nm), **b** 5p ^{2}P$_{3/2}$ → 9s
^{2}S$_{1/2}$ (246.8-nm), **c** 5p ^{2}P$_{3/2}$
→ 5s5p^2 ^{4}P$_{5/2}$ (283.7-nm)
and **d** 5p ^{2}P$_{1/2}$ → 6s ^{2}S$_{1/2}$
(410-nm) transitions after
ToF corrections. Fits by a
Voigt profile are shown by
the solid lines. © 2020
American Physical Society

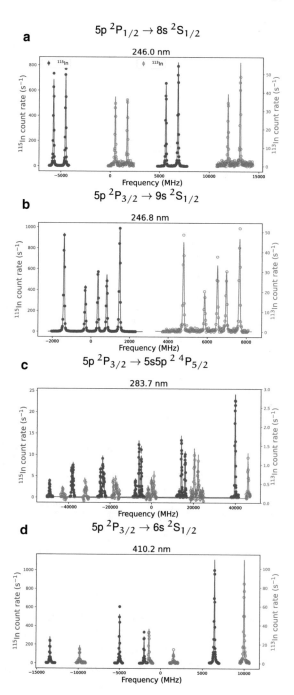

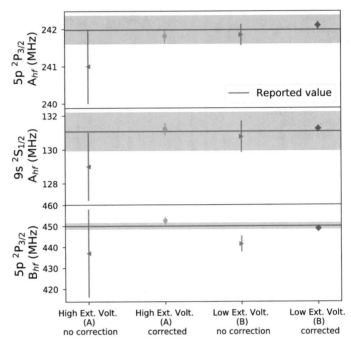

Fig. 5.16 Comparison of extracted hyperfine structure constants for the 5p $^2P_{3/2}$ → 9s $^2S_{1/2}$ (246.8-nm) transition with the low and high-voltage extraction configurations, as well for parameters extracted with and without using the ToF corrections described in Sect. 5.3. High Ext. Volt. corresponds to 5 V/mm and Low Ext. Volt. to 0.02 V/mm from the simulations shown in Fig. 5.10. © 2020 American Physical Society

Table 5.5 Isotope shift values extracted from measurements with the 5p $^2P_{1/2}$ → 8s $^2S_{1/2}$ (246.0-nm), 5p $^2P_{3/2}$ → 9s $^2S_{1/2}$ (246.8-nm) and 5p $^2P_{1/2}$ → 6s $^2S_{1/2}$ (410-nm) transitions. Statistical and systematic uncertainties are shown by the curly and square brackets respectively

λ (nm)	Transition	$\delta\nu^{115,113}$ (MHz) literature [Ref]	This work
246.0	5p $^2P_{1/2}$ → 8s $^2S_{1/2}$		−268(2)[4]
246.8	5p $^2P_{3/2}$ → 9s $^2S_{1/2}$		−270(2)[6]
410.2	5p $^2P_{1/2}$ → 6s $^2S_{1/2}$	−258(3) [38]	−260(3)[5]

precision by an order of magnitude. The measured isotope shift values for the 5p $^2P_{3/2}$ → 9s $^2S_{1/2}$ (246.8-nm), 5p $^2P_{1/2}$ → 8s $^2S_{1/2}$ (246.0-nm) and 5p $^2P_{1/2}$ → 6s $^2S_{1/2}$ (410-nm) transitions are displayed in Table 5.5 and agree with the available previous measurement of the 5p $^2P_{1/2}$ → 6s $^2S_{1/2}$ (410-nm) transition (Table 5.5).

5.5 Benchmarking Coupled-Cluster Calculations

Coupled-cluster calculations in the singles and doubles approximation, with a perturbative correction to triple excitations ('CCSD(T)') and without the correction ('CCSD'), were performed by [39] to determine the atomic parameters needed to precisely extract the nuclear quadrupole moments, Q_S, of indium. Previously the most precise measurement of ^{115}In was made with indium halide molecules using quantum-chemical calculations to determine the quadrupole coupling constant [40]. A comparison to literature with the Q_S values determined in this work is later made in Table 5.8.

The atomic components of the hyperfine structure constants in Eq. 2.26, 2.27 were calculated in the forms A_{hf}/g_I and B_{hf}/Q, independent of the nuclear system, for the nuclear gyromagnetic factor g_I and the nuclear quadrupole moment Q_S respectively. In Table 5.6 a comparison is made between calculations of the A_{hf} and B_{hf} constants using DHF alone and with the CCSD(T) approach [39] (see Sect. 2.3.2) for the six atomic states measured in this work.

The contribution from each of these corrections is also shown in Table 5.6. The difference between DHF and CCSD values makes apparent the varying importance of electron-correlation effects for each state, where the DHF value has a larger value for only the $5p\ ^2P_{3/2}$ state, with it being smaller for all other states. The correction from

Table 5.6 Comparison of calculated atomic parameters A_{hf}/g_I and B_{hf}/Q using DHF and CCSD methods. Corrections to the CCSD approach from the Breit interaction, QED effects and triples are also stated. Calculations by Sahoo [8, 41]

	$5p\ ^2P_{1/2}$	$5p\ ^2P_{3/2}$	$6s\ ^2S$	$7s\ ^2S$	$8s\ ^2S$	$9s\ ^2S$
A_{hf}/g_I (MHz) values						
DHF	1454.81	218.78	800.55	274.17	127.36	72.56
Values from CCSD method						
DC	1841.52	207.96	1321.11	415.95	187.41	106.88
Breit	−2.67	0.10	1.61	0.64	0.30	0.15
QED	−3.11	0.78	−5.81	−2.11	−1.08	−0.78
Triples	11.27	−3.68	19.07	7.87	2.59	1.46
Uncer. ±	20.0	8.0	30.0	15.0	8.0	3.0
B_{hf}/Q (MHz/b) values						
DHF		419.83				
Values from CCSD method						
DC		581.61				
Breit		−2.27				
QED		0.04				
Triples		−3.78				
Uncer. ±		4.0				

triples was found to be large compared to the Breit or QED corrections. Higher-order excitations were not taken into account but are expected to have a smaller contribution [39].

A comparison with the A_{hf} constants measured in this work is made in Table 5.7 where the atomic factors, A_{hf}/g_I, are combined with high-precision nuclear magnetic dipole measurements of ^{115}In and ^{113}In by NMR, with values of $\mu = +5.5408(2)\ \mu_N$ [42] and $\mu = +5.5289(2)\ \mu_n$ [43] respectively.

A comparison with the B_{hf} constants cannot be made as atomic-model-independent measurements of the nuclear electric quadrupole moment exist, which would be needed to evaluate B_{hf} from the calculated electric field gradient factor, B_{hf}/Q. There is 1.5σ agreement of the theoretical CCSD values with the values measured in this work and those from literature for both ^{115}In and ^{113}In. The inclusion of triples with the CCSD(T) calculation increased the accuracy of the A_{hf} values, from 15 MHz to 10 MHz for the 5p ^{2}P$_{3/2}$ state for example, while a similar 1.5σ level of agreement was maintained. Improved agreement with experiment and a reduction in calculated uncertainties may be achieved by a full treatment of the triple excitations.

SD calculations, which only consider linear terms in the CCSD method [8] were previously made by [44] and are also compared in Table 5.7. Significant differences are present between the SD and CCSD calculations, highlighting the importance of non-linear terms included with the RCC method in accurately including electron correlations.

CCSD calculations of the A_{hf} values for these atomic states were originally reported in [12], and are shown in the literature column of Table 5.7. In addition to an increased accuracy of our extracted Q_S values, the improvement of these calculations also benefits atomic EDM measurement experiments. The indium atom has been shown to have a large EDM enhancement factor, and its relatively smaller size, compared to the present best limit for a paramagnetic atom EDM from Tl [45], would allow for improved accuracy of the atomic factors. The limit to the accuracy of the atomic factors in these studies is from the error in the weak-interaction Hamiltonians used. These errors are evaluated from the off-diagonal matrix elements of the magnetic dipole interaction Hamiltonian, which are proportional to $\sqrt{A_{hf}}$ of the corresponding states in the matrix elements [12, 46, 47]. The 5p ^{2}P$_{1/2}$ and 6s to 9s ^{2}S$_{1/2}$ states listed in Table 5.7 were found to give the leading contributions.

The A_{hf} values of the 8s ^{2}S$_{1/2}$ and 9s ^{2}S$_{1/2}$ states were previously not measured by experiment until this work. Agreement with experiment improved with the values from the new CCSD(T) calculations, with the most significant difference for the 9s ^{2}S$_{1/2}$ state. Compared to the previous calculations [12], the biggest differences are the inclusion of higher-order relativistic terms, non-linear terms in the RCC hyperfine interaction operator and the use of quadratic (QTO) instead of Gaussian type orbitals (GTO), which are more appropriate for high-lying s states [39, 48]. Measurement of the 5s5p^2 ^{4}P$_{5/2}$ A_{hf} and B_{hf} constants for ^{113}In were also made for the first time in this work with the 283-nm (5p ^{2}P$_{3/2}$ → 5s5p^2 ^{4}P$_{5/2}$) transition. However due to the lower atomic physics incentive and the type of orbitals used, the hyperfine structure constants were not calculated for the 5s5p^2 ^{4}P$_{5/2}$ A_{hf} [8].

Table 5.7 Hyperfine structure constants of ^{115}In and ^{113}In extracted from measurements with the 246.0-nm ($5p\,^2P_{1/2} \to 8s\,^2S_{1/2}$), 246.8-nm ($5p\,^2P_{3/2} \to 9s\,^2S_{1/2}$), 283-nm ($5p\,^2P_{3/2} \to 5s5p^2\,^4P_{5/2}$) and 410-nm ($5p\,^2P_{1/2} \to 6s\,^2S_{1/2}$) transitions. Statistical and systematic uncertainties are shown by the first and second brackets respectively. NMR measurements values of ^{115}In and ^{113}In with $\mu = +5.5289(2)\,\mu_N$ [43] and $\mu = +5.5408(2)\,\mu_N$ [42] were used to compare to A_{hf}/g_I values

Level	A_{hf} (MHz)						B_{hf} (MHz)	
	Theory				Experiment		Experiment	
	Literature		Sahoo [8, 41]					
	SD [44]	CCSD [12]	CCSD*	CCSD(T)†	Literature [Ref]	This work	Literature [Ref]	This work
^{115}In								
$5p\,^2P_{1/2}$	2306	2256(30)	2260(30)	2274(25)	2281.9501(4) [49]	2282.04(45)(53)		
$5p\,^2P_{3/2}$			257(15)	253(10)	242.1647(3) [49]	241.98(12)(26)	449.545(3) [49]	450(1)(0.5)
$6s\,^2S_{1/2}$	1812	1611(50)	1621(50)	1645(37)	1685.3(6) [32]	1684.75(76)(39)		
$7s\,^2S_{1/2}$	544.5	516(30)	510(30)	520(19)	541.0(3) [36]			
$8s\,^2S_{1/2}$	240.8	234(20)	230(20)	233(10)		243.85(24)(19)		
$9s\,^2S_{1/2}$	128.1	106(10)	131(10)	136(6)		131.07(98)(15)		
$5s5p^2\,^4P_{5/2}$					3654(13) [36]	3696(1)(2)	-644(13) [36]	-660(30)(3)
					3689(20) [37]		-638(50) [37]	
^{113}In								
$5p\,^2P_{1/2}$			2255(30)	2270(25)	2277.0860(4) [50]	2277.07(34)(77)		
$5p\,^2P_{3/2}$			256(15)	252(10)	241.6409(4) [50]	241.75(25)(21)	443.414(4) [50]	442(3)(1)
$6s\,^2S_{1/2}$			1618(50)	1642(37)	1681.8(8) [51]	1682.47(98)(76)		
$7s\,^2S_{1/2}$			509(30)	519(19)				
$8s\,^2S_{1/2}$			229(20)	231(10)		243.49(24)(23)		
$9s\,^2S_{1/2}$			130(10)	136(6)		130.87(94)(38)		
$5s5p^2\,^4P_{5/2}$						3690(2)(3)		-558(35)(5)

*—The uncertainties of the CCSD calculated values are estimated due to neglected triple excitation configurations

†—The CCSD(T) uncertainties were calculated from comparison to a lower-order basis function [8]

The literature A_{hf} value of ^{115}In for the 5s5p^2 ^{4}P$_{5/2}$ state measured by [36] differs from our measured value by 3σ. A value reported by [37] agrees within 1σ of both our value and that of [36] due to a large uncertainty. Although disagreement between our value is 7 MHz compared to 35 MHz from the [36] value. As A_{hf} was also measured for ^{113}In, a comparison can be made with the ratio of magnetic moments, reported by NMR measurements as 1.00215(5) [42, 43]. This gives a ratio of 1.0003(16) for our measurement and 0.990(4) for the measurement by [36]. Our reported value is within 1.2σ rather than 3σ, giving support to our reported values for the 5s5p^2 ^{4}P$_{5/2}$ state.

5.6   Nuclear Moments

The calculated atomic parameters of Table 5.6 could then be applied to the measured hyperfine structure constants to determine the nuclear moments of ^{115}In and ^{113}In, the results are presented in Table 5.8. The A_{hf} constant from the 5p ^{2}P$_{1/2}$ state was used to determine the magnetic dipole moment as it had the smallest uncertainty and the B_{hf} constant from the 5p ^{2}P$_{3/2}$ state was used to determine the nuclear electric quadrupole moment. The atomic parameters were $A_{hf}(^2P_{1/2})/g_I = 1847(20)$ MHz and $B_{hf}(^2P_{3/2})/Q_S = 576(4)$ MHz/b respectively.

The determined magnetic dipole moments also agree well with the high-precision values from NMR [42, 43], giving confidence in the atomic parameters used and in the accuracy of the Q_S values evaluated in Table 5.8 and Q_S values of the neutron-rich isotopes determined in Chap. 7.

The Q_S values reported here for ^{115}In and ^{113}In agree within uncertainty with the previously reported most precise values in literature [40], and improves upon the precision by 1 mb and 30 mb respectively. Using literature values for B_{hf} (indicated by † in Table 5.8) further improves the precision by 2 and 6 mb.

Table 5.8 Nuclear magnetic dipole moments and nuclear electric quadrupole moments determined using the atomic parameters $A_{hf}(^2P_{1/2})/g_I = 1847(20)$ MHz and $B_{hf}(^2P_{3/2})/Q_S = 576(4)$ MHz/b

	^{113}In		^{115}In	
	Literature [Ref]	This work	Literature [Ref]	This work
μ (μ_N)	5.5289(2) [43]	5.548(62)	5.5408(2) [42]	5.560(62)
Q_S (b)	0.80(4) [32]	0.767(11)*	0.83(10) [52]	0.781(7)*
		0.770(5)†	0.770(8) [40]	0.780(5)†

*—with the measured hyperfine structure constants in Table 5.7
†—using literature A_{hf} and B_{hf} values [49, 50]

5.7 Atomic State Lifetime Measurement

The lifetimes of the upper states of the 246.0-nm (5p $^2P_{1/2}$ → 8s $^2S_{1/2}$), 246.8-nm (5p $^2P_{3/2}$ → 9s $^2S_{1/2}$), and 410-nm (5p $^2P_{1/2}$ → 6s $^2S_{1/2}$) transitions were measured by delaying the Q-switch of the final ionization 1064-nm laser and measuring the change in ionization rate on the MCP detector. The fits to the de-excitation curves are shown in Fig. 5.17a, b, c). No transitions are known to exist which would populate states above the 37268 cm^{-1} threshold for ionization by the 1064-nm light from de-excitation of the $^2S_{1/2}$ upper states. The integrated count rate over the leftmost hyperfine structure peak ($F = 6$ → 5) was measured for each 1064-nm light delay value. Typical measurement times of 10 min per Q-switch delay measurement were used, the measurement time for each scan was used to normalise the total count rate. The pulse duration of the 1064-nm laser is quoted as 10 ns, while the pulse duration FWHM of the IS laser was measured to be a 50 ns. The timing precision of the pulse delay generator is quoted to a level of 500 ps. No jitter in the relative timing of the laser pulses from lasers was observed at a 1 ns level using a photodiode (which will include Q-switch and pulse delay generator timing jitter) at the end of the beamline, this was therefore used as the upper limit to the level of relative timing precision.

The remaining count-rate errors were likely from the fluctuations in atom beam current, laser-atom overlap and atom-laser overlap. Measurements of the resonant count rate at the same delay values at the start and end of the lifetime measurements were used to estimate an uncertainty in the resonant count rate to account for longer term drifts in atom beam current and laser intensity. The results from these measurements are shown in Table 5.9 and are compared to literature values.

As shown in Fig. 5.17a, b, c), it was found that a term for a constant background rate needed to be included, which is shown by the fit in blue (the fit in red is without). These fits correspond to the values in Table 5.17. While the magnitudes of the lifetimes are correct, the values for the all three upper states are overestimated. In Fig. 5.17a, b) a slight curvature is observed with respect to the log scale of the count rate.

From the remaining unaccounted sources of error, the most likely explanation for the overestimation of the lifetime is due to variations in laser-atom overlap as a function of Q-switch delay. This could explained if the resonantly ionized ions have a different ion transmission to the detector depending on the position in the interaction region they are created, or if the atom-laser overlap is not perfectly uniform.

When gating on the quartiles of the ToF distribution for the detected bunch, for the $5s^29s\ ^2S_{1/2}$ upper state, lifetimes of 521(37) and 373(20) ns were extracted for the upper and lower quartiles respectively. There is therefore a dependence on the ToF of the detected ions. This source of error is included in the square brackets in Table 5.17, which brings the $5s^26s\ ^2S_{1/2}$ state lifetime into 1σ agreement with literature and reduces the disagreement for the $5s^29s\ ^2S_{1/2}$ and $5s^28s\ ^2S_{1/2}$ states (2σ, 3σ).

As the detected bunch represents only the ionized component of the full atom beam this only gives a partial explanation for the overestimation of lifetime, but suggests that the laser-atom overlap does indeed need to be accounted for. A dedicated study of the measurement systematics in a well known system would be needed to confirm

Fig. 5.17 Log scale of the count rate used to determine the lifetimes of the **a** $5s^2 8s\ ^2S_{1/2}$ (40 636.98 cm^{-1}), **b** $5s^2 9s\ ^2S_{1/2}$ (42 719.02 cm^{-1}) and **c** $5s^2 6s\ ^2S_{1/2}$ (24 372.957 cm^{-1}) states determined by fitting of an exponential to the Q-switch delay of the second step laser (the ionization step). Fits including a constant background rate are show in blue, those without in red

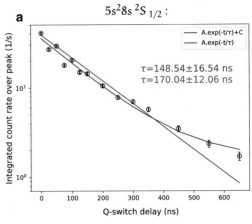

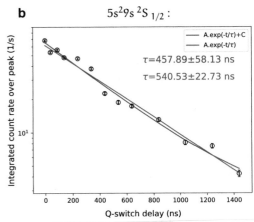

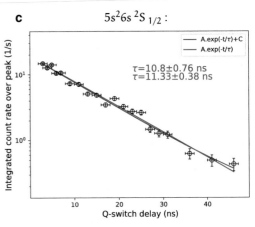

Table 5.9 Lifetimes of the upper states extracted from an experiment on stable Indium isotopes using the ablation source. Values extracted from a fit of exponential decay to the Q-switch delay of the second step laser (the ionization step). A systematic error from the gated quartile of the ToF is included in square brackets, as discussed in the text

Transition	Upper state	Lifetime measured (ns)	Lifetime literature (ns) [ref.]
246.8 nm	$5s^2 9s\ ^2S\ _{1/2}$	457(58)[147]	104(12)[53]
246.0 nm	$5s^2 8s\ ^2S\ _{1/2}$	149(17)[40]	55(6)[53]
410 nm	$5s^2 6s\ ^2S\ _{1/2}$	10.8(8)[30]	7.0(3) [54]

the source of the error in the overestimation of the lifetimes, in order to make use of a multi-step laser ionization schemes for reliable lifetime measurements in the future with this setup.

5.8 Isotopic Abundance

The isotopic abundance of ^{113}In relative to ^{115}In was additionally measured in this work as the relative resonance ionization rate of the masses was measured in several scans. As mentioned in Sect. 5.3.2, the isotope mass selectivity came from the selectivity of the laser ionization, due to the difference in absolute hyperfine transition frequencies of ^{113}In versus ^{115}In. Using the ToF alone, the mass resolution is poor, $m/\Delta m \approx 1$, this is indicated in red in Fig. 5.7.

The relative peak intensity between ^{113}In and ^{115}In for the same hyperfine split atomic transitions in a spectrum was used to determine the relative isotopic abundance for each scan. This has the advantage over integrating the total counts in each region that fluctuations which change the measured count rate when on resonance, are included in the amplitude error of the fitted resonances. As shown in Fig. 5.4, the level of fluctuation in beam intensity reduced with increasing time from the start of the ablation laser. Although all of the measurements were made after around an hour of starting the ablation laser, determining the relative isotopic abundance in this way from fitting the resonance peaks of the hyperfine structure gave an uncertainty estimation from the remaining fluctuations in atom-beam intensity or laser power.

Figure 5.18a, b) show the variation in ^{113}In abundance with the number of hyperfine scans whilst Table 5.10 gives the average value for each transition. The combined average of the relative isotopic abundance of ^{113}In extracted from the 5p $^2P_{3/2} \rightarrow$ 9s $^2S_{1/2}$ (246.8-nm) and 5p $^2P_{1/2} \rightarrow$ 8s $^2S_{1/2}$ (246.0-nm) transition measurements agree well with the natural abundance of ^{113}In [55].

See Fig. 5.19 for the variation in the isotopic abundance of ^{113}In extracted over 15 hyperfine structure measurements when integrating the total count rate. The distribution has a very small uncertainty (<0.1% per scan) due to the large number of counts, despite a standard deviation of 2% isotopic abundance. In comparison with

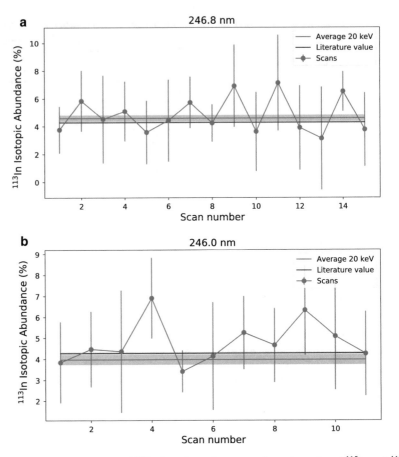

Fig. 5.18 Natural abundances of ^{113}In from hyperfine spectra intensity ratios of ^{115}In and ^{113}In over many scans of the **a** 5p $^2P_{3/2}$ → 9s $^2S_{1/2}$ (246.8-nm) **b** 5p $^2P_{1/2}$ → 8s $^2S_{1/2}$ (246.0-nm) transition

Table 5.10 The natural abundance of ^{113}In extracted from hyperfine transition intensity ratios of ^{115}In and ^{113}In, compared to literature values [55]

	^{113}In abundance (%)
246.8 nm	4.59(22)
246.0 nm	3.99(24)
Average	**4.30(16)**
Literature	4.28(5)[55]

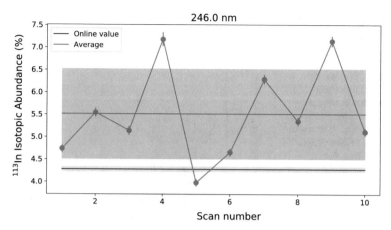

Fig. 5.19 Natural percent abundance of ^{113}In extracted from summing counts with the 5p ^{2}P$_{3/2}$ → 9s ^{2}S$_{1/2}$ (246.8-nm) transition. This demonstrates the non-statistical scatter when integrating the counts over a hyperfine spectrum without accounting for resonance rate fluctuations during the scan

the approach to extract the isotopic abundances shown in Fig. 5.18a, b), integrating the counts takes no account of fluctuations during a scan, which explains the non-statistical scatter seen in the figure.

5.9 Atomic Populations Following Charge Exchange

The relative populations of the 5p ^{2}P$_{1/2}$ and 5p ^{2}P$_{3/2}$ states in indium were also measured with the ablation ion source, at a beam energy of 20 keV. Following neutralisation with the reactions of the type

$$\text{In}^+(0) + \text{Na}(0) \rightarrow \text{In}(k) + \text{Na}^+(0) + \Delta E \, , \tag{5.3}$$

and after 120 cm of ion flight, the relative ionization rate of the 5p ^{2}P$_{1/2}$ → 8s ^{2}S$_{1/2}$ (246.8-nm) and 5p ^{2}P$_{3/2}$ → 9s ^{2}S$_{1/2}$ (246.0-nm) transitions were measured, giving the relative populations of the 5p ^{2}P$_{1/2}$ and 5p ^{2}P$_{3/2}$ states. The ionization laser was delayed by 20 ns from the resonant excitation step to avoid lineshape distortions from the high-power pulsed light [56] but to maintain the laser ionization efficiency. The close proximity in wavelength of the two transitions enabled rapid switching between them (<10 min), maintaining near identical laser beam properties between two measurements with each transition. Reproducible spacial overlap of the atoms and ions was ensured by alignment irises and Faraday cups through which both ion and laser beam transport was maximised. The 532-nm ablation laser remained active while the resonant laser was being tuned to the other transition to minimise the change in beam intensity from oxidation or cooling of the target.

Fig. 5.20 Comparison of experimental to simulated values of the relative atomic populations of atomic indium 5p $^2P_{1/2}$ and 5p $^2P_{3/2}$ states following neutralisation of indium ions. The relative population of the simulated values was normalised to the count rate for the 5p $^2P_{1/2}$ state

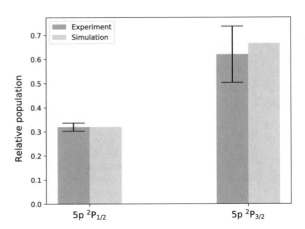

The integrated count rate of a hyperfine spectrum from each transition was used to give the proportion of atomic populations in the lower states (5p $^2P_{1/2}$ and 5p $^2P_{3/2}$). Normalisation of the atom beam intensity and uncertainty from fluctuations was taken into account using the background rate between the hyperfine structure peaks in the spectra. It was not possible to use the same procedure as in Sect. 5.8 as this cannot be performed between different transitions, additionally in Sect. 5.8 the background rate was not measured reliably as only a short scan region around a single hyperfine transition was measured. The background count rate was measured with a relative uncertainty of ~18 and 5% for the 5p $^2P_{3/2} \rightarrow$ 9s $^2S_{1/2}$ (246.8-nm) and 5p $^2P_{1/2} \rightarrow$ 8s $^2S_{1/2}$ (246.0-nm) transitions respectively, the reduced uncertainty for the 5p $^2P_{1/2} \rightarrow$ 8s $^2S_{1/2}$ (246.0-nm) transition was due to the much larger scan region off resonance due to the larger splitting (A_{hf}) of the hyperfine structure.

The atomic level populations measured in this work, in addition to the results of the simulation (described in Sect. 2.4) are displayed in Fig. 5.20 and Table 5.11. The ratio of the ionization rates from the states 5p $^2P_{1/2}$ and 5p $^2P_{3/2}$ was used for comparison to simulation, as the populations are relative (to the 5p $^2P_{1/2}$ state in this case).

The experimental values agree within 1σ with those simulated, although the experimental uncertainty is large. This gave confidence in extending the simulations to other atomic systems as well to a beam energy of $T_B = 40$ keV, which was later used for the measurements of the neutron-rich indium isotopes. The results were also compared against available experimental data for fluorine [57] and nickel [58] in Table 5.11.[3]

---

[3] Reprinted from Vernon, A. R. et al. Simulation of the relative atomic populations of elements $1 \leq Z \leq 89$ following charge exchange tested with collinear resonance ionization spectroscopy of indium. Spectrochim. Acta Part B At. Spectrosc. 153, 61–83, Copyright (2019) with permission from Elsevier.

Table 5.11 Comparison to available data of relative atomic populations after charge exchange

	T_B (keV)	l_{fl} (cm)	State	Sim. Rel. Pop.	Exp.	Sim. Rel. Pop. [58]
In	20	120	$5p\,^2P_{1/2}$	0.319	*	
In	20	120	$5p\,^2P_{3/2}$	0.665	0.619(116)	
F	40.8	190	$3s\,^4P_{5/2}$	0.19	0.083(4) [57]	
Ni	29.85	40	$4s\,^3D_3$	0.130	*	
Ni	29.85	40	$4s\,^3D_2$	0.107	0.114(10) [58]	0.1138(100)
Ni	29.85	40	$4s\,^3D_1$	0.06	0.00(1) [58]	0.083(1)
Ni	29.85	40	$4s\,^1D_2$	0.02435	0.0229(46) [58]	0.0354(1)

*—indicates the atomic populations were taken relative to this state i.e. divided by the count rate for the transition from the state

The normalisation of the beam current using the atomic fraction was the largest uncertainty in the reported values. In future work this uncertainty could be significantly reduced by using a particle detector to detect the neutral current, from the portion of the beam which is not ionized and deflected, rather than the collisional re-ionization rate, which is purposely suppressed with the CRIS technique. This could be achieved using an ITO coated glass plate [59, 60], which is able to detect current from atom impact and has photoelectric effect suppression making it insensitive to the laser pulses collinear to the neutral fraction. Normalisation of the laser intensities using a photodiode or charge-coupled detector could be used to normalise fluctuations in laser power density (if not in a saturated regime), allowing this source of error to be reduced. Combined with these planned improvements, the experimental setup would be very effective for measuring the relative atomic populations.

The results of the simulations compared to the relative atomic population measurements of this thesis and previous work is displayed in Table 5.11. To the authors knowledge, this table shows all data for atomic populations measured in the energy range $T_B = 1$–100 keV to date.

The magnitudes of the simulated atomic populations largely agree, but are outside $> 1\sigma$ of the experimental values in two cases. The simulated final population for the $3s\,^4P_{5/2}$ state in fluorine is a factor of two greater than the measured value, while the simulated initial population is 0.079, which would agree with the experimental value. This could be explained by i) overestimation in the transition rates available for fluorine in literature ii) the contribution from high-lying states could be reduced if collisionally ionized before decay into the state of interest (see Sect. 6.3.3 for a discussion of collisional ionization from high-lying states). iii) an error in the preferential ionization [61] factor (for re-ionization with a helium cell) used by [57] in the evaluation of the population from the measurement iv) or neglected experimental uncertainties.

The simulated value was 0.06 for the 4s ^{3}D$_1$ state in nickel, however no population was observed in the values reported by [58] for nickel, despite the strong (A_{ij} = $1.2 \times 10^8 \ s^{-1}$) 4s ^{3}D$_1$ → 4p ^{3}P$_0$ transition being used. One explanation could be inaccurate branching ratios of the transitions, de-exciting the initial population, which would reduce the population of the 4s ^{3}D$_1$ state.

As the calculation method has been shown to be useful in predicting relative populations in these cases, the calculations have been extended to all elements from $Z = 1$ to $Z = 89$ with available atomic data, for incident ions $A^+(0)$ upon potassium and sodium vapours $B(0)$

$$A^+(0) + B(0) \rightarrow A(k) + B^+(0) , \tag{5.4}$$

where both the vapour and ion beam are again assumed to be in their ground states. The results are shown in Tables B.1 and B.2 for potassium and sodium vapours respectively, in the appendix.

While simulations were performed for $1 \le Z \le 89$ for completeness, for atomic numbers $Z \le 3$ at $T_B = 40$ keV, their velocities are over the boundary of the assumed intermediate velocity region. While for heavier systems relativistic calculations become more important [62] and therefore validity of the above approach reduces. The reliability has not been tested for very light or heavy systems and is therefore unknown, this is also true for systems where very little atomic data is known. These cases are marked in Tables B.1 and B.2. The results are displayed for the initial population distribution at 0 cm from charge exchange, and for the final population distribution following 120 cm of atom flight, for beam energies of 5 keV and 40 keV. The five most highly populated states are listed in these tables. In lighter systems such as hydrogen or helium, much experimental cross-section data and cross-section calculations are available [63]. However as a cross section needs to be determined for every state in order to determine the initial relative populations, the approximate approach used in this work (see Sect. 2.4.2) is less computationally expensive. In the case of systems with very few known transition probabilities, the initial populations prior to the decay part of the simulations are quoted but their final population following decay are not. Their atomic analogues can be used as an indicator for their expected relative atomic populations. The relative populations in Tables B.1 and B.2 were normalised by the total cross section for all of the charge-exchange reactions, these values are displayed in Table B.3, the accuracy of these values is highly dependent upon the completeness of available atomic data however.

5.10   Conclusions

The practicality of a collinear resonance ionization beamline in tandem with an ablation ion source has been demonstrated using indium as an example case. Although the idea to combine the two systems is simple, it is the first time to the authors knowledge that this has been performed, along with the time-of-flight corrections, and it allows for a significantly reduced complexity of a collinear resonance ionization setup.

A minimum in the energy spread of the ion beam produced with the ablation source was found with a low extraction electric field configuration (0.01 V/mm). This was directly observable in the lineshapes of the measured hyperfine spectra due to the high-resolution of the injection-seeded laser (20 MHz).

However, the use of a 500 ps timing resolution ion time-of-flight data acquisition allowed the reconstruction of high-resolution spectra in low- or high-field extraction configurations (down to 60(8) MHz with low field). Although the energy spread was a hindrance for measurement of atomic parameters, when combined with time-of-flight information further insight was gained into laser-plasma interactions [64, 65] as a function of time. The systematics of the laser ablation plasma process measured from hyperfine structures will be explored in future work at CRIS. In addition, the time-of-flight information may be of use for measurements of other systems with large energy spread, such as atomic or molecular systems which are insufficiently cooled.

Measurements of the the hyperfine-structure constants of the $5s^2 8s \ ^2S_{1/2}$ and $5s^2 9s \ ^2S_{1/2}$ states were made for the first time using this setup. These measurements highlight the potential for precision atomic physics studies which this low-complexity apparatus opens up. The A_{hf} constants of the $5s^2 8s \ ^2S_{1/2}$ and $5s^2 9s \ ^2S_{1/2}$ states are sensitive parameters for the benchmarking relativistic coupled-cluster calculations [12]. The corrections that were needed for improved agreement between the CCSD(T) values and measured A_{hf} constants gave confidence in the accuracy of the electric-field gradient parameter B_{hf}/Q_S for the 5p $^2P_{3/2}$ state, which was later used for the studies of radioactive indium isotopes.

In addition, the field shifts for the 5p $^2P_{1/2} \rightarrow 8s \ ^2S_{1/2}$ (246.0 nm) and 5p $^2P_{3/2} \rightarrow$ 9s $^2S_{1/2}$ (246.8-nm) transitions, ^{113}In / ^{115}In isotopic abundance ratio and lifetimes of the $5s^2 6s \ ^2S_{1/2}$, $5s^2 8s \ 2S \ 1/2$ and $5s^2 9s \ ^2S_{1/2}$ states were measured.

The relative isotopic abundance of ^{113}In to ^{115}In obtained, 4.30(16)%, is in agreement with the literature value. However the lifetime measurements were largely overestimated for the three upper states measured, a 30% variation in the extracted lifetime values with the ToF of the detected ions suggests the problem may be due to inhomogeneous laser-atom overlap. Further studies of the systematics are needed before this setup could be reliably used for atomic lifetime measurement. Lifetime measurements would provide complementary information for the testing of underlying atomic theories by comparison to calculated lifetimes [66].

Simulations were performed to predict the relative atomic populations of the atomic population distribution of indium ions, following neutralisation by a sodium vapour, at a beam energy of 20 keV and 40 keV This included calculating the cross sections for neutralisation of the atomic states in indium using a semi-classical impact-parameter approach [67], and then the simulation of the population redistribution by de-excitation over 120 cm. Measurements were made of the relative atomic populations of indium neutralised with sodium at 20 keV, also using the ablation source. The population of the two lowest-lying states in indium, 5p $^2P_{1/2}$ and 5p $^2P_{3/2}$, were extracted and are in agreement with the simulated population.

The simulations were then extended to predict the relative atomic population with ions of atomic numbers $1 \leq Z \leq 89$ neutralised by sodium and potassium vapours (the results are shown in Tables B.2 and B.1 respectively). Beam energies of 5 keV and 40 keV were used, as this covers the energy range of typical values used for CRIS and collinear laser spectroscopy experiments. A comparison was made with literature measurements for fluorine and nickel. The reproduction of the overall population distribution was found to be good, however two states were measured to have values outside of experimental uncertainty; possible explanations for the discrepancies were given. These results will be of use for future collinear laser spectroscopy experiments, as prediction of atomic populations is vital for choosing or developing an efficient laser scheme, while experimental data remains extremely sparse. The verification of the relative population of the $5p\ ^2P_{1/2}$ and $5p\ ^2P_{3/2}$ states at 20 keV gave confidence that using sodium to neutralise the neutron-rich indium isotopes from ISOLDE at 40 keV would give the highest relative population of the $5p\ ^2P_{3/2}$ state. The feasibility of the ablation source with the CRIS technique for measuring relative atomic populations has been demonstrated. The principle improvement to the setup which would allow for extraction of populations with higher accuracy is the simultaneous recording of the atom beam current, for normalisation of the total beam intensity.

The $5p\ ^2P_{3/2} \rightarrow 9s\ ^2S_{1/2}$ (246.8-nm) transition was found to have a compact hyperfine structure, desirable for the online measurements detailed in Sect. 6. The total hyperfine structure of <4 GHz gave A_{hf} and B_{hf} with uncertainties of $<1\%$ and $<5\%$ respectively, compared to <14 GHz for the 283.7 nm transition. This was important when step sizes as small as 6 MHz/min were required to gather enough statistics to resolve the structure of the most exotic isotopes.

The potential compatibility of the ablation source with a solid target of any material allows the study of many atomic and molecular species without the need for cooling and bunching [68]. This in principle allows highly reactive elements such as oxygen and carbon to be studied. Producing bunched oxygen, carbon or fluorine [57] beams presents a challenge at ISOLDE due to the ready formation of molecules in both the ISOLDE targets and the ISCOOL cooler buncher. An additional advantage of using CRIS is the population of high lying meta-stable states in charge exchange, creating accessible transitions to laser spectroscopy which were not possible previously due to the deep-UV lasers or two-photon transitions [69, 70] that would be needed to excite from the ground state. This is highlighted in the relative populations simulated in Tables B.2 and B.1 of the appendices. Laser spectroscopy studies of these elements would be important for fundamental nuclear structure [71], and also environmental science [72], and in general the low background, high efficiency, high selectivity of a CRIS setup is well suited to molecular spectroscopy or trace analysis such as for geophysical dating [73–76].

The ablation source setup also has potential applications in future exploratory studies of molecular systems such as HfF or BaF, which are highly sensitive to parameters relevant to fundamental physics [77, 78]. In systems such as these, precise measurement of transitions is important for both verification of quantum-chemical calculations and for developing laser cooling schemes [78, 79]. The source can support studies of radioactive molecules, which are produced in the ISOLDE target [80], but for which very little is known about the isotopic dependence of their hyperfine structure [8, 81]. Recent high-resolution measurements [8] of transitions in RaF molecules produced at ISOLDE show that CRIS the technique is indeed suitable for molecular measurements. The measurement of radioactive molecules will provide a new source of valuable nuclear structure information for which this ablation ion source may provide valuable supporting studies.

The ablation source and CRIS setup is well suited to exploratory studies as compared with directly in traps, as limitations due to space-charge effects (10^{12} ions/laser pulse is possible) [82], refilling trap time [83] and distribution of states [84] are avoided with collinear laser spectroscopy. Additionally a separation of typically $<10\mu s$ between neutralisation and resonance excitation allows access to metastable states not possible with traps.

References

1. Lynch KM et al (2018) Phys Rev C 97(2):1. ISSN 24699993
2. Wraith C et al (2017) Phys Lett B 771:385, 2017. ISSN 0370-2693. https://www.sciencedirect.com/science/article/pii/S0370269317304483
3. De Groote R et al (2017) Phys Rev C 96(4):1. ISSN 24699993
4. Farooq-Smith GJ et al (2016) Phys Rev C 94(5):054305. ISSN 2469-9985. http://link.aps.org/doi/10.1103/PhysRevC.94.054305
5. Flanagan KT et al (2009) Phys Rev Lett 103(14):142501. ISSN 0031-9007. http://link.aps.org/doi/10.1103/PhysRevLett.103.142501 https://link.aps.org/doi/10.1103/PhysRevLett.103.142501
6. de Groote RP et al (2017) Hyperfine Interact 238(1):5. ISSN 15729540
7. Aratari R (1988) Nuclear Instrum Methods Phys Res Sect B: Beam Interact Mater Atoms 34(4):493. ISSN 0168-583X. https://www.sciencedirect.com/science/article/pii/0168583X88901565
8. Garcia Ruiz RF et al (2018) In CERN-INTC-2018- P546. https://cds.cern.ch/record/2299760/files/INTC-P-546.pdf
9. Kumaki M et al (2016) Rev Sci Instrum 87(2):02A921. ISSN 0034-6748. http://aip.scitation.org/doi/10.1063/1.4939781
10. Olmschenk et al S (2017) Appl Phys B 123(4):99. ISSN 0946-2171. http://link.springer.com/10.1007/s00340-017-6683-1
11. Chichkov BN et al (1996) Appl Phys A Mater Sci Process 63(2):109. ISSN 0947-8396. http://link.springer.com/10.1007/BF01567637
12. Sahoo BK et al (2011) Phys Rev A 030502(84):5
13. Pospelov et al M (2005) Ann Phys 318(1):119. ISSN 00034916. http://www.sciencedirect.com/science/article/pii/S0003491605000539
14. Jamba DM (1969) Rev Sci Instrum, 40(8):1072. ISSN 0034-6748. http://aip.scitation.org/doi/10.1063/1.1684155
15. Hasted JB (1972) Physics of atomic collisions, 2nd edn. American Elsevier. https://bibdata.princeton.edu/bibliographic/1917772
16. Menzinger M (1969) Rev Sci Instrum 40(1):102. ISSN 00346748. http://scitation.aip.org/content/aip/journal/rsi/40/1/10.1063/1.1683697
17. Conrads H et al (2000) Plasma Sourc Sci Technol 9(4):441. ISSN 0963-0252. http://stacks.iop.org/0963-0252/9/i=4/a=301?key=crossref.4fd3f4207ddcb8c7bfc482edfbe7b293

18. Kirchner R (1990) Nuclear Instrum Methods Phys Res Sect A: Acceler, Spectrom, Detect Assoc Equipm 292(2):203. ISSN 0168-9002. https://www.sciencedirect.com/science/article/pii/016890029090377I
19. Lettry J et al (2003) Nuclear Instrum Methods Phys Res Sect B: Beam Interact Mater Atoms 204:363. ISSN 0168-583X. https://www.sciencedirect.com/science/article/pii/S0168583X02019675
20. Blais J et al (1976) Int J Mass Spectrom Ion Phys 20(2-3):329. ISSN 00207381. https://www.sciencedirect.com/science/article/pii/0020738176801593
21. Liebl H (2008) Applied charged particle optics. Springer, Berlin. http://link.springer.com/10.1007/978-3-540-71925-0
22. Vernon A (2019) Nuclear instruments and methods in physics research section b: beam interactions with materials and atoms, EMIS2018:submitted
23. Gao F et al (2014) AIP advances 4(2):027118. ISSN 2158-3226. http://aip.scitation.org/doi/10.1063/1.4866983
24. Torrisi L et al (2003) Appl Surf Sci 210:262. http://web3.le.infn.it/sub-web/leas/Pubblicazioni/COMPARISONOFNANOSECONDLASERABLATION.pdf
25. Maul J et al (2005) Phys Rev B 71(4):045428. ISSN 1098-0121. https://link.aps.org/doi/10.1103/PhysRevB.71.045428
26. Anoop KK et al (2015) Article J Appl Phys. https://www.researchgate.net/publication/272815013
27. Wang X et al (2014) Spectrochim Acta Part B: Atomic Spectr 99:101. ISSN 0584-8547. https://www.sciencedirect.com/science/article/pii/S0584854714001220
28. Liénard E et al (2005) Nuclear Instrum Methods Phys Res Sect A: Acceler, Spectr, Detect Assoc Equip 551(2-3):375. ISSN 0168-9002. https://www.sciencedirect.com/science/article/pii/S0168900205013811
29. Saquilayan GQ et al (2017) Proceedings 1869:40004. https://doi.org/10.1063/1.4995705 https://doi.org/10.1063/1.4995780
30. Auciello O et al (1993) J Vacuum Sci Technol A: Vacuum, Surf, Films 11(1):267. ISSN 0734-2101. http://avs.scitation.org/doi/10.1116/1.578715
31. Sonnenschein V et al (2017) Laser Phys 27(8):085701. ISSN 1054-660X. http://stacks.iop.org/1555-6611/27/i=8/a=085701?key=crossref.ae700ab1aa5d94924db40c8ed77febe7
32. Eberz J et al (1987) Nuclear Phys A 464(1):9. ISSN 03759474. http://linkinghub.elsevier.com/retrieve/pii/0375947487904192
33. Stancik AL et al (2008) Vibr Spectr 47:66
34. C. AB (2019) COMSOL Multiphysics. www.comsol.com
35. Zhigilei LV et al (1998) J Phys Chem B. https://pubs.acs.org/doi/abs/10.1021/jp9733781
36. George S et al (1990) J Opt Soc Am B 7(3):249. ISSN 0740-3224. https://www.osapublishing.org/abstract.cfm?URI=josab-7-3-249
37. Zhao YY et al (1986) Zeitschrift fur Physik D Atoms, Mol Clust 3(4):365. ISSN 0178-7683. http://link.springer.com/10.1007/BF01437193
38. Neijzen J et al (1980) Phys B+C 98(3):235. ISSN 0378-4363. https://www.sciencedirect.com/science/article/pii/0378436380900832
39. Sahoo BK (2018) Private communication. Breit and QED effects, CCSD(T) calculations for indium including higher order relativistic effects
40. van Stralen JNP et al (2002) J Chem Phys 117(7):3103. ISSN 0021-9606. http://aip.scitation.org/doi/10.1063/1.1492799
41. Sahoo BK (2018) Private communication. Relativistic coupled cluster calculations for atomic states in indium
42. Flynn CP et al (1960) Proc Phys Soc 76(2):301. ISSN 0370-1328. http://stacks.iop.org/0370-1328/76/i=2/a=415?key=crossref.f3515cb119eec3150e31da423e1fdb38
43. Rice M et al (1957) Phys Rev 106(5):953. ISSN 0031-899X. https://link.aps.org/doi/10.1103/PhysRev.106.953
44. Safronova UI et al (2007) Phys Rev A - Atom, Mol, Opt Phys 76(2):1. ISSN 10502947. https://link.aps.org/doi/10.1103/PhysRevA.76.022501

45. Regan BC et al (2002) Phys Rev Lett 88(7):071805. ISSN 0031-9007. https://link.aps.org/doi/10.1103/PhysRevLett.88.071805
46. Nataraj HS et al (2011) Phys Rev Lett 106(20):1. ISSN 00319007
47. Nataraj HS et al (2008) Phys Rev Lett 101(3):033002. ISSN 0031-9007. https://link.aps.org/doi/10.1103/PhysRevLett.101.033002
48. Sahoo BK (2016) Phys Rev A 93(2):022503. ISSN 2469-9926. https://link.aps.org/doi/10.1103/PhysRevA.93.022503
49. Eck et al TG (1957) Phys Rev 106(5):958. ISSN 0031-899X. https://link.aps.org/doi/10.1103/PhysRev.106.958
50. Eck TG et al (1957) Phys Rev 106(5):954. ISSN 0031-899X. https://link.aps.org/doi/10.1103/PhysRev.106.954
51. Jackson DA (1981). https://www.sciencedirect.com/science/article/pii/0378436381901509 https://ac.els-cdn.com/0378436381901509/1-s2.0-0378436381901509-main.pdf?_tid=bcc9c7a3-8c03-4aca-bc08-c255df0285da&acdnat=1540981628_1a413161eb822aa965578b36fc9c568b
52. Batty C et al (1981) Nuclear Phys A 355(2):383. ISSN 0375-9474. https://www.sciencedirect.com/science/article/pii/0375947481905340
53. Jönsson G et al (1983) Phys Rev A 27(6):2930. ISSN 0556-2791. https://link.aps.org/doi/10.1103/PhysRevA.27.2930
54. Lubowiecka TDT (1978) Acta Physica Polonica A 54:369
55. Haynes WM (2015) CRC handbook of chemistry and physics : a ready-reference book of chemical and physical data, 96th edn. CRC Press, Boulder, Colorado, USA
56. de Groote RP et al (2017) Phys Rev A 95(3):032502. ISSN 2469-9926. http://link.aps.org/doi/10.1103/PhysRevA.95.032502
57. Levy CDP et al (2007) Nuclear Instrum Methods Phys Res Sect A: Acceler, Spectr, Detect Assoc Equipm 580(3):1571. ISSN 01689002
58. Ryder C et al (2015) Spectrochim Acta Part B: Atom Spectr 113:16. ISSN 05848547. http://linkinghub.elsevier.com/retrieve/pii/S0584854715001962
59. Hanstorp D (1992) Meas Sci Technol 3(5):523. ISSN 09570233. http://stacks.iop.org/0957-0233/3/i=5/a=013
60. Rothe S et al (2017) J Phys G: Nuclear Particle Phys 44(10):104003. ISSN 0954-3899. http://stacks.iop.org/0954-3899/44/i=10/a=104003?key=crossref.853522bbb0fcafbb7110d8ae82ed31a8
61. Firsov OB (1959) J Exptl Theoret Phys (USSR) 36(36):1517
62. Stöhlker T et al (1998) Phys Rev A 58(3):2043. ISSN 1050-2947. https://link.aps.org/doi/10.1103/PhysRevA.58.2043
63. Ghanbari-Adivi E et al (2014) Phys Script 89(10):105402. ISSN 0031-8949. http://stacks.iop.org/1402-4896/89/i=10/a=105402?key=crossref.f4cb20c2206b4ade6932a9d9b53fc82d
64. Hegelich et al BM (2006) Nature 439(7075):441. ISSN 0028-0836. http://www.nature.com/articles/nature04400
65. Amoruso S et al (1999) J Phys B: Atomic, Mol Opt Phys Topic Rev J Phys B: At Mol Opt Phys, 32:131, 1999. http://iopscience.iop.org/article/10.1088/0953-4075/32/14/201/pdf
66. Patterson BM et al (2015) Phys Rev A 91(1):012506. ISSN 1050-2947. https://link.aps.org/doi/10.1103/PhysRevA.91.012506
67. Rapp D et al (1962) The Journal of Chemical Physics, 37(11):2631, 1962. ISSN 00219606. http://scitation.aip.org/content/aip/journal/jcp/37/11/10.1063/1.1733066
68. Mané E et al (2009) Eur Phys J A, 42(3):503, 2009. ISSN 14346001
69. Chrysalidis K (2018) First demonstration of Doppler-free two-photon in-source laser spectroscopy at ISOLDE-RILIS EMIS 2018. Technical report, CERN. https://indico.cern.ch/event/616127/contributions/3006973/attachments/1718328/2807166/6_EMIS_2018_KC.pdf
70. Mirza MY et al (1978) Proc R Soc Lond. Ser A, Math Phys Sci 364(1717):255. https://www.jstor.org/stable/pdf/79762.pdf?refreqid=excelsior%3Ae4940d0c0a786dd9ee4234f22ba8187d

71. Kumar A et al (2017) Phys Rev Lett 118(26):262502. ISSN 0031-9007. https://arxiv.org/pdf/1705.05409.pdf http://link.aps.org/doi/10.1103/PhysRevLett.118.262502
72. Wood BJ et al (2010) Nature 465(7299):767. ISSN 0028-0836. http://www.nature.com/articles/nature09072
73. Wang et al Z (2013) Nature 499(7458):328. ISSN 0028-0836. http://www.nature.com/articles/nature12285
74. Naeraa et al T (2012) Nature 485(7400):627. ISSN 0028-0836. http://www.nature.com/articles/nature11140
75. Kaiser J et al (2012) Surf Sci Rep 67(11-12):233. ISSN 0167-5729. https://www.sciencedirect.com/science/article/pii/S0167572912000416#f0005
76. Kudriavtsev YA et al (1982) Appl Phys B Photophys Laser Chem 29(3):219. ISSN 0721-7269. http://link.springer.com/10.1007/BF00688671
77. Cairncross WB et al (2017) Phys Rev Lett 119(15):153001. ISSN 0031-9007. https://link.aps.org/doi/10.1103/PhysRevLett.119.153001
78. Kozyryev et al I (2017) Phys Rev Lett 119(13):133002. ISSN 0031-9007. https://link.aps.org/doi/10.1103/PhysRevLett.119.133002
79. Isaev TA et al (2010) Phys Rev A 82(5):052521. ISSN 1050-2947. https://link.aps.org/doi/10.1103/PhysRevA.82.052521
80. Kronberger M et al (2013) Nuclear Instrum Methods Phys Res Sect B: Beam Interact Mater Atoms 317:438. ISSN 0168-583X. https://www.sciencedirect.com/science/article/pii/S0168583X13008422
81. Aldegunde J et al (2018) Phys Rev A 97(4):042505. ISSN 2469-9926. https://link.aps.org/doi/10.1103/PhysRevA.97.042505
82. Loeb HW (2005) Plasma Phys Control Fus 47(12B):B565. ISSN 0741-3335. http://stacks.iop.org/0741-3335/47/i=12B/a=S41?key=crossref.0d0c23f6304e5089b17428e813547f4d
83. Porobić T et al (2015) Nuclear Instrum Methods Phys Res Sect A: Accelerat, Spectrom, Detect Assoc Equipm 785:153. ISSN 0168-9002. https://www.sciencedirect.com/science/article/pii/S0168900215002673
84. Ni K-K et al (2014) J Mol Spectr 300:12. ISSN 0022-2852. https://www.sciencedirect.com/science/article/pii/S0022285214000332#f0005

Chapter 6
Analysis of Neutron-Rich Indium Studied with Laser Spectroscopy

This chapter describes the analysis procedure implemented in order to take the measured ion count rates and resonant step laser frequencies and extract hyperfine structure parameters from fitting of the wavenumber-binned spectra, in addition to the process used to quantify and correcting for known sources of experimental error.

6.1 Data Sorting Procedure

The measured ion counts from the MCP detector and wavenumbers from the wavemeter were saved in the form of timestamped arrays during the experiments. To construct the hyperfine spectra the two arrays were first correlated in time by backfilling [1] the wavenumber data for each measured count. This was necessary as the data acquisition was not event-by-event, with the wavelength and count rate being recorded at different rates, therefore the timestamps were not synchronised. This is a valid approximation as the time difference between wavelength measurements and ion counts timestamps were at the <5 ms level while the maximum laser frequency scan speed used was at the level of 6 kHz per ms on average, where the finer laser step sizes were a function of the unknown digital-to-analog converter precision used inside the seed laser (see Sect. 4.1.4). Corrections to the measured laser frequencies were also made at this stage, to account for drifts in the accuracy of the wavemeter and beam energy, T_B, which was used to convert the frequencies to the atomic rest frame (see Sect. 6.1.1.2).

The counts were then binned by laser frequency, using the same bin width as the laser frequency step used to scan the laser. This was typically 6 MHz per step (one second per step) for high-resolution scans using the M-Squared SolsTiS for the injection-seeded laser (Sect. 4.1.4). For each laser frequency step taken to collect a spectrum, the frequency was taken as a 'setpoint' with a feedback loop between

A. R. Vernon, *Collinear Resonance Ionization Spectroscopy of Neutron-Rich Indium Isotopes*, Springer Theses, https://doi.org/10.1007/978-3-030-54189-7_6

the wavemeter and laser that adjusted the laser optics (mainly the etalon) until the frequency was reached. Only when the setpoint was reached would the next step be taken. As the amount of time this takes was not uniform, it creates small differences in the actual scan speed. The non-uniform time per step was corrected for by converting the total number of counts in each frequency bin into a count rate by using the total time spent in each frequency bin.

In some cases multiple scans were combined in order to enhance the statistics of spectra for a given isotope, or to combine partial scans of a hyperfine spectrum which had a large hyperfine splitting. This was performed similarly by working out the time spent at each wavelength and the corresponding number of counts for the separate scans before binning the data to construct a final spectra with a weighted count rate.

Figures 6.1 and 6.2 show a compilation of scans from the even- and odd-mass indium isotopes measured during the experiment. A summary of the number of scans taken at each mass with the 246.8 nm ($5p\,^2P_{3/2} \rightarrow 9s\,^2S_{1/2}$) and 246.0 nm ($5p\,^2P_{1/2} \rightarrow 8s\,^2S_{1/2}$) transitions is shown in Table 6.1.

6.1.1   Frequency Corrections

There were two principle sources of error in the determination of hyperfine transition frequencies during these experiments. The largest source of error (up to 150 MHz) was due to drifting of the wavelengths measured by the wavemeter due to imperfect calibration, which was attributed to thermal fluctuations in the laboratory temperature. For this reason the frequency of a diode laser was used as a reference to correct for such drifts in the wavemeter, which was locked to a single frequency, see Sect. 4.3.

The second largest source of error was due to changes in the ISCOOL extraction voltage, which determines the ion beam energy and therefore gives the level of accuracy the spectra could be Doppler-shift corrected to. If the ISCOOL voltage changes are on a scale longer than the duration of a scan (typically 30 min) then the hyperfine structure parameters, A_{hf} and B_{hf}, will be unaffected by the change (the error from shorter time-scale voltage drifts are assessed in Sect. 6.3). But as the centre of gravity of the hyperfine structure will be shifted due to an incorrect Doppler shift, the isotope shift values will be in error. A change in 1 V at mass 115 corresponds to a centroid shift of 14 MHz at a beam energy of 40 keV.

In order to quantify and correct for these and other sources of error in the transition frequencies measured, the hyperfine spectrum of ^{115}In was measured before and after each measurement of the neutron-rich isotopes (or at most two measurements). The independent measurements of the hyperfine spectra of ^{115}In allowed for the accuracy of the neutron-rich isotopes measurements to be validated and allowed the quantification of systematic correlations over time.

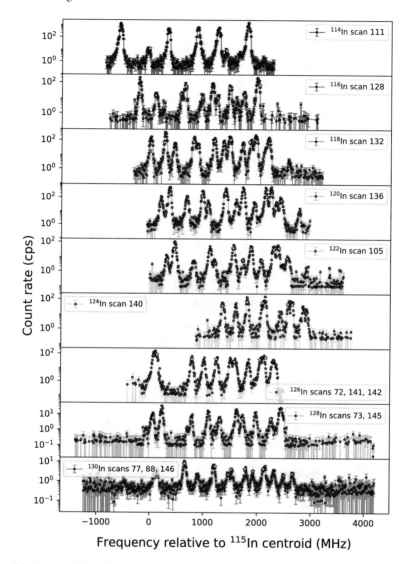

Fig. 6.1 A compilation of hyperfine spectra at each measured odd indium isotope mass. Each isotope has at least one isomer, also visible in the spectra and indicated by the coloured star makers

6.1.1.1 Wavemeter Correction

The accuracy of the recorded wavelength by the wavemeter used in these experiments (Sect. 4.3) varies with temperature and pressure in its environment. By using a fibre-optic switch box and a diode reference laser with a high long-term stability (<1 MHz specified [2]), the correction relation

$$\frac{\Delta f_1}{f_1} = \frac{\Delta f_2}{f_2} , \tag{6.1}$$

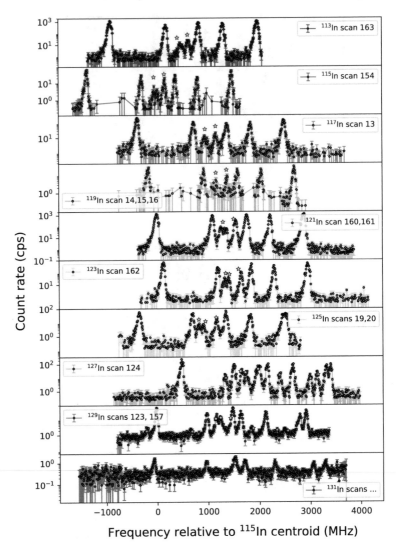

Fig. 6.2 A compilation of hyperfine spectra at each measured even indium isotope mass. Each isotope has at least one isomer, also visible in the spectra and indicated by the coloured star makers

can be used (within ± 200 nm of the calibration wavelength [2]), for this Fizaeu type wavemeter [3]. Allowing the drift in resonant laser frequency to be corrected with the recorded diode frequency, as both measurements are made with the same interferometer [4] in the wavemeter unit. The reference diode laser was locked to a 795 nm hyperfine transition in rubidium [5] (see Sect. 4.3). This transition was then used as a reference frequency, f_2, allowing any change in this frequency, Δf_2,

Table 6.1 Number of full scan ranges completed with the 5p $^2P_{3/2} \rightarrow$ 9s $^2S_{1/2}$ (246.8 nm) and 5p $^2P_{1/2} \rightarrow$ 8s $^2S_{1/2}$ (246.0 nm) transitions. The 'minutes' (m) listed in these tables indicates the total amount of time spend scanning the laser frequencies over the hyperfine structure at each mass

Mass	Minutes (# scans) with transition 246.8 nm	246.0 nm	Mass	Minutes (# scans) with transition 246.8 nm	246.0 nm
113	32 min. (2)	24 min. (1)	123	28 min. (1)	–
114	101 min. (2)	61 min. (1)	124	26 min. (part.)	–
115	958 min. (38)	243 min. (9)	125	50 min. (2)	–
116	91 min. (2)	27 min. (1)	126	87 min. (1)	–
117	16 min. (1)	43 min. (1)	127	132 min. (2)	137 min. (2)
118	124 min. (2)	36 min. (1)	128	129 min. (2)	34 min. (1)
119	20 min. (1)	–	129	431 min. (4)	182 min. (2)
120	75 min. (2)	49 min. (1)	130	254 min. (3)	94 min. (1)
121	27 min. (1)	–	131	908 min. (6)	343 min. (1)
122	147 min. (2)	48 min. (part.)			

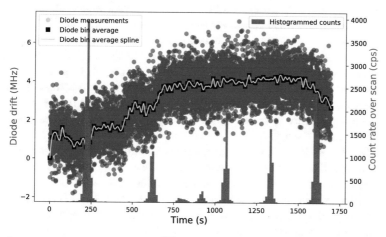

Fig. 6.3 Example hyperfine spectrum of ^{115}In binned in time compared to the drift in wavemeter calibration measured using the diode reference laser A binning and averaging of 10 s was used

recorded by the wavemeter to be used to work out the corresponding correction needed, Δf_1, for the measured injection seeded laser frequency, f_1.

Figure 6.3 shows the recorded drift in the calibration of the wavemeter as shown by the diode laser during the course of a measurement of the hyperfine structure of ^{115}In. The scatter in the individual measurements of the wavelength of the diode laser can be seen, compared to an average of 10 s along with the corresponding time-binned hyperfine spectrum. Figure 6.4 shows the count rate observed as a function of

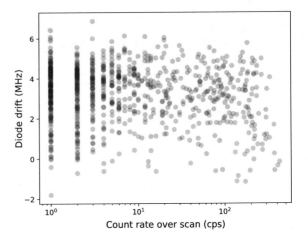

Fig. 6.4 Count rate plotted as a function of the closest wavemeter frequency of the diode reference laser. No correlation between change in observed diode frequency and count rate was observed

the closest measurement in time of the diode reference laser. The wavelength sample rate was 100 ms compared to the number of detected ions recorded every 10 ms. No correlation between the resonance count rate and the drift in the frequency of the diode laser was seen, indicating that this shorter time-scale scatter did not correlate with a change in resonance count rate over the course of the scan. The correlation with the extracted ^{115}In centroids with the longer time-scale drift is discussed below.

The source of this scatter is not known, but may be due to thermal or electrical noise, however it is below the level of the specified precision of the wavemeter used (Sect. 4.3). Figure 6.5a, b shows the effect of using the recorded drift in the diode frequency as a correction to the recorded resonant laser frequency, and the corresponding decrease in scatter of the extracted centre-of-gravity frequencies of the ^{115}In reference spectra.

The 5p ^{2}P$_{3/2}$ → 9s ^{2}S$_{1/2}$ (246.8 nm) transition exhibited significantly more deviation in the recorded ^{115}In centroids compared to the 5p ^{2}P$_{1/2}$ → 8s ^{2}S$_{1/2}$ (246.0 nm) transition, although there was a more stable period for around 20 h.

The pulsed fundamental light produced by the injection-seeded Ti:Sa (Sect. 4.1.4) was fibre coupled directly to the wavemeter before frequency tripling. The one possible explanation is due to the use of a multi-mode fibre-optic switch box, used to allow measurement of multiple laser frequencies with the single-channel wavemeter. This may have allowed multiple spatial modes to exist for measurement by the wavemeter, and may explain the short time-scale fluctuations seen in Fig. 6.5a as a function of time for the 5p ^{2}P$_{3/2}$ → 9s ^{2}S$_{1/2}$ (246.8 nm) transition, if the dominant mode was fluctuating.

Differences between the fibre coupling for the 738 and 740.4 nm fundamental light, for the 5p ^{2}P$_{3/2}$ → 9s ^{2}S$_{1/2}$ (246.8 nm) and 5p ^{2}P$_{1/2}$ → 8s ^{2}S$_{1/2}$ (246.0 nm) transitions respectively, may have had lead to a difference in the extent to which this effect was present.

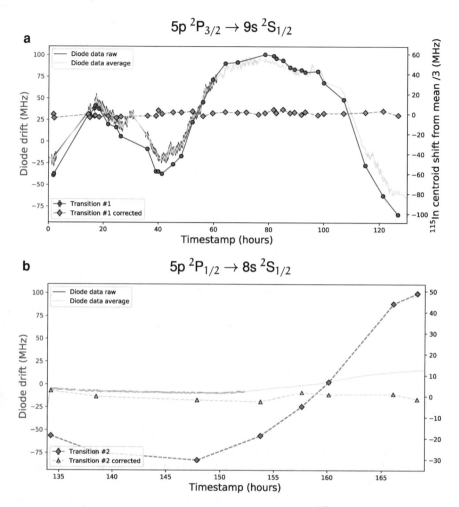

Fig. 6.5 The effect of correction to the centre-of-gravity shift of the ^{115}In hyperfine spectra, using the wavemeter correction based on the recorded reference diode wavelength. Shown for the: **a** Transition #1: 5p ^{2}P$_{3/2}$ → 9s ^{2}S$_{1/2}$ and **b** Transition #2: 5p ^{2}P$_{1/2}$ → 8s ^{2}S$_{1/2}$. See Fig. 6.6 for the correlation plots before correction. The centroid frequencies are divided by 3 to show the frequency drifts on the scale of the fundamental light which was measured

Additionally, differences in the fibre coupling may have lead to differences in the energy deposition of the fundamental pulsed light into the wavemeter, which could have caused a higher level of thermal fluctuations for one than the other.

The correlation between the diode wavelength drift and the measured ^{115}In centroid drift was very strong in both cases, as shown in Fig. 6.6. After applying the correction the spread in the ^{115}In centroids was reduced from ±50 MHz to ±2.5 MHz for the 5p ^{2}P$_{3/2}$ → 9s ^{2}S$_{1/2}$ (246.8 nm) transition, and reduced from ±15 MHz to ±1 MHz for the 5p ^{2}P$_{1/2}$ → 8s ^{2}S$_{1/2}$ (246.0 nm) transition. This was a significant

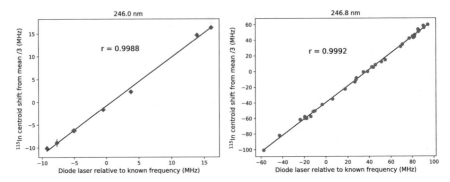

Fig. 6.6 Correlation plots of the ^{115}In hyperfine spectra centroids for the $5p\,^2P_{1/2} \to 8s\,^2S_{1/2}$ (246.0 nm) and $5p\,^2P_{3/2} \to 9s\,^2S_{1/2}$ (246.8 nm) transitions versus the reference diode laser frequency recorded by the wavemeter. r is the correlation coefficient

correction to the accuracy of the extracted isotope shifts, however the shorter time-scale scatter observed during measurements with the $5p\,^2P_{3/2} \to 9s\,^2S_{1/2}$ (246.8 nm) transition may still have been the principle limitation to the accuracy of both isotope shift and hyperfine constants later reported in this work.

6.1.1.2   ISCOOL Voltage Doppler Correction

Corrections due to drifts in the ISCOOL extraction voltage were applied *following* the wavemeter correction described above. The magnitude of change in the ^{115}In centroid frequencies due to the wavemeter drift (± 50 MHz) would have obscured the correlation between the ISCOOL voltage and the ^{115}In centroid frequencies, which are on the order of a few MHz in the 113–131 atomic mass region of indium.

The ISCOOL extraction voltage was recorded every few seconds during the experiment. The measurement was made by a resistive voltage divider connected to an analog-to-digital voltmeter. It was therefore not possible to decouple real drifts in voltage from changes in the resistance of the voltage divider, which is sensitive to temperature and RF fields. A more recently constructed and calibrated precision high-voltage divider borrowed from the University of Mainz [6] was used to calibrate the voltage divider just before the measurements of the experiment began. This gave a calibration fit which was used as a correction to the recorded ISCOOL extraction voltage values in this analysis as $V_{correct} = 0.99877(1) \times V_{measured} + 0.027(22)$.

Figure 6.7 shows the ^{115}In centroids following the wavemeter correction alongside the recorded ISCOOL extraction voltage. It can be seen from this that the remaining scatter in the ^{115}In centroids is of the order expected for the changes in ISCOOL extraction voltage for the $5p\,^2P_{1/2} \to 8s\,^2S_{1/2}$ (246.0 nm) transition (2 MHz), while a larger scatter remains for the $5p\,^2P_{3/2} \to 9s\,^2S_{1/2}$ (246.8 nm) transition (10 MHz). The correlation coefficients [7] for the ^{115}In centroids with scatter in voltage recorded during the measurements with the $5p\,^2P_{3/2} \to 9s\,^2S_{1/2}$ (246.8 nm) and $5p\,^2P_{1/2} \to 8s$

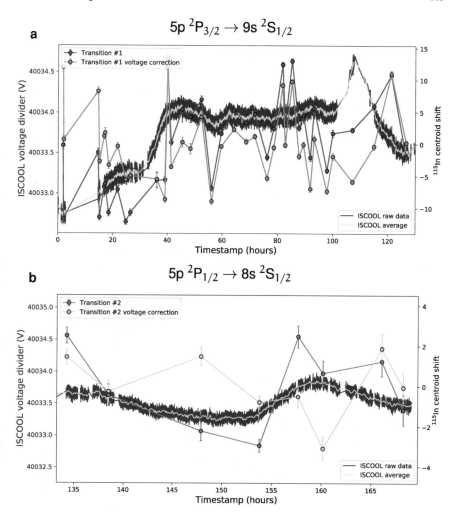

Fig. 6.7 The change on the centre of gravity of the ^{115}In hyperfine spectra using a Doppler correction based on the recorded ISCOOL extraction voltage. The correlation coefficients were $r = 0.8$ and $r = 0.1$ for the **a** Transition #1: $5p\,^2P_{3/2} \to 9s\,^2S_{1/2}$ (246.8 nm) and **b** Transition #2: $5p\,^2P_{1/2} \to 8s\,^2S_{1/2}$ (246.0 nm) transitions respectively

$^2S_{1/2}$ (246.0 nm) transitions were determined to be r=0.1 and r=0.8 respectively. For the $5p\,^2P_{3/2} \to 9s\,^2S_{1/2}$ (246.8 nm) transition a remaining systematic error dominates, possibly from the smaller time scale drift in the wavemeter calibration, seen in the diode reference wavelength mentioned above. While for the $5p\,^2P_{1/2} \to 8s\,^2S_{1/2}$ (246.0 nm) transition there is a clearer correlation, hence the Doppler correction was applied to the spectra using the ISCOOL voltage as a function of time.

The difference in masses for the isomer states compared to the ground states of isotopes in indium also contribute a Doppler shift. In this work the isotope masses were taken from the 2016 atomic mass evaluation [8], and isomer excitation energies from the Nuclear Data Sheets [9, 10]. Of the measured states the majority have excitation energies <500 keV, which corresponds to a Doppler shift in the hfs centroid shift of <1.5 MHz. For the the $^{21}/_2{}^-$ and $^{23}/_2{}^-$ states in ^{127}In and ^{129}In, with excitation energies of 1863 keV and 1630 keV respectively, this shift becomes more significant as the excitation corresponding to shifts of 8 MHz and 7 MHz in the centroid of the spectra respectively. Excitation energy Doppler corrections were performed by re-binning and fitting the spectra with the corrected additional mass for the excited isomer state.

6.2   Hyperfine Structure Fitting

Once binned by frequency the hyperfine spectra were fitted using a model of the hyperfine structure based on the known or tentative nuclear spin I, and the known electronic upper J_u and lower J_l states for the transition. The fitting was performed in Python with χ^2 minimisation of the model to a binned spectrum using the package LMFIT [11, 12].

Equation 2.25 was used for the upper and lower electronic angular momenta, J_{upper}, J_{lower}, to give the transition frequency between hyperfine levels ν_{hf} as

$$\nu_{hf} = \nu_{cog} + \alpha_{upper} A_{upper} + \beta_{upper} B_{upper} - \alpha_{lower} A_{lower} - \beta_{lower} B_{lower} \quad (6.2)$$

the centroids of each hyperfine structure peak in the structure as well as the centre of gravity of the structure ν_{cog}, was fitted using this relationship which allowed the hyperfine constants A_{hf}, B_{hf} and isotope (or isomer) shift values to be determined. The relative peak intensities were initially calculated using Eq. 2.31, but left as a free parameters in the fitting as the polarisation of the light was not recorded, although it was expected to not change after the harmonic-generation crystal. The simultaneous fitting of hyperfine models to separate spectra was also performed in some cases with the aid of linked hyperfine models in the package SATLAS [13]. The A_{hf} ratio obtained between the transitions from the reference isotope ^{115}In was used for this purpose.

The additional constraint from linked model fitting was also valuable in assigning peaks to the structures measured with the 5p ^{2}P$_{3/2}$ → 9s ^{2}S$_{1/2}$ (246.8 nm) transition, for which the spectra contained hyperfine structures from up to three nuclear states. Whereas the 5p ^{2}P$_{1/2}$ → 8s ^{2}S$_{1/2}$ (246.0 nm) transition had a maximum of 4 peaks per state, with a large separation due to the large A_{hf} value.

For many of the indium isotopes measured with the 5p ^{2}P$_{3/2}$ → 9s ^{2}S$_{1/2}$ (246.8 nm) transition, the rightmost hyperfine structure peaks were unresolved (the $F = 4 \rightarrow F = 5$ and $F = 3 \rightarrow F = 4$ transitions) due to the limited linewidth of the injection-seeded laser (34 MHz [14]), for example in Fig. 5.14. For this reason and the relatively small A_{hf} and B_{hf} values for the transition (see Table 6.4), the correlation map shown

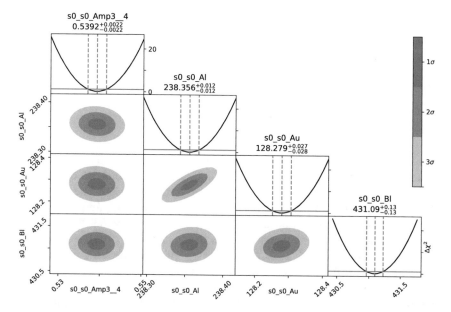

Fig. 6.8 Correlation map to determine if any of the fitted HFS parameters (Al, Au, Bl) are correlated with each other or with the amplitude of one of the overlapping HFS peaks (Amp3_4). No significant correlations are found. Mapped for unresolved atomic transitions using a scan of ^{115}In with the 246.8 nm transition. Map generated using the Python package SATLAS [13]

in Fig. 6.8 was generated to determine if this overlap would cause any correlation in the parameters by their uncertainties. Only a slight correlation between the A_{lower} and A_{upper} parameters was observed (at the level of <0.02% inside 1σ uncertainty of the parameters), showing that the values can compensate for each other to a small but negligible extent.

6.2.1 Lineshape

The lineshape used to fit all of the hyperfine structure resonances this work was a Voigt profile [15], resulting from the convolution of a Gaussian profile with a Lorentzian profile. Where the Gaussian component is attributed to inhomogeneous broadening such as Doppler broadening and where the Lorentzian component is from the natural lineshape of the transition and homogeneous broadening such as power broadening [16]. For computational convenience the Voigt profile was evaluated as

$$V(x, \sigma, \gamma) = \frac{\text{Re}[w(z)]}{\sigma\sqrt{2\pi}} \, , \tag{6.3}$$

where $\mathrm{Re}[w(z)]$ is the real part of the Faddeeva function [17]

$$w(z) = e^{-z^2} \, \mathrm{erf}(-iz) \, , \tag{6.4}$$

where $\mathrm{erf}()$ is the error function and the parameter

$$z = \frac{x + i\gamma}{\sigma\sqrt{2}} \, , \tag{6.5}$$

includes the Lorentzian and Gaussian FWHM contributions, γ and σ, respectively. The relative contribution to the FWHM of the Gaussian or Lorentzian components was left as a free parameter for the peaks. These parameters were shared between all of the components of the hyperfine structures in a given spectrum, as they are independent of the nuclear structure or atomic transition. Figure 6.9 shows the relative Gaussian and Lorentzian contributions to the linewidth over the experiment plotted alongside the χ_r^2 of the fit to the peaks. While the average total linewidth remains around 47(9) MHz, the relative Gaussian to Lorentzian component changes and the increasing Gaussian component is correlated with an increased χ_r^2. This can be attributed to the use of too high of an atom beam current, which has the combined effect of magnifying any present fluctuations in laser or atom beam intensity beyond the level of the statistical scatter. In addition to causing saturation of the data-acquisition electronics at the highest count rate component of the peak. This causes the Gaussian contribution to increase to compensate, as for a given amplitude a Gaussian profile has a larger FWHM than a Lorentzian profile. The χ_r^2 values would increase from this and from the higher statistics scatter in the resonant count

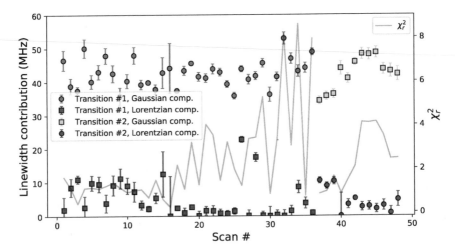

Fig. 6.9 The relative Gaussian and Lorentzian contributions with scan number for Transition #1: $5p\,^2P_{3/2} \rightarrow 9s\,^2S_{1/2}$ (246.8 nm) and Transition #2: $5p\,^2P_{1/2} \rightarrow 8s\,^2S_{1/2}$ (246.0 nm). The Lorentzian component disappears for increased χ_r^2 of the fitted peaks

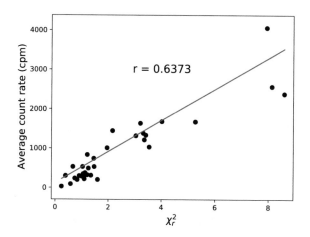

Fig. 6.10 Correlation between the average count rate over the ^{115}In reference scans and the increase in the χ_r^2 of the hyperfine structure fit. With a correlation coefficient of r = 0.64

rate. Figure 6.10 displays the correlation between the χ_r^2 of the fitted profiles and the average count rate for the ^{115}In reference measurements, which gives support to this hypothesis.

A peak asymmetry was observed in the hyperfine spectra obtained with both the $5p\,^2P_{1/2} \rightarrow 8s\,^2S_{1/2}$ (246.0 nm) and $5p\,^2P_{3/2} \rightarrow 9s\,^2S_{1/2}$ (246.8 nm) transitions in this work (shown in Fig. 6.11 for the 246.0 nm transition). An understanding of the source of the asymmetry is important for determining the centroid of the measured hyperfine transition accurately. Not modelling the asymmetry correctly can introduce an offset which would reduce the accuracy of the isotope shift values, if the asymmetry is not constant between the isotope measurements.

From the population simulations (see Sect. 2.4), the asymmetry is predicted to be partially accounted for by the population of the $5p\,^2P_{1/2}$ and $5p\,^2P_{3/2}$ states by the decay of intermediate states following neutralisation. This is due to the additional energy E^* used to populate these intermediate states compared to the direct population of the $5p\,^2P_{1/2}$ or $5p\,^2P_{3/2}$ states. The required energy is taken from the kinetic energy of the beam, causing a Doppler shift in the resonance frequency for a fraction of the atomic beam and therefore a contribution to the resonance lineshape offset from the nominal beam energy.

To show this, the beam energy before the reaction $T_{B,I}$ with that after the reaction $T_{B,II}$ with an energy difference in IP of $\Delta E = I_A - I_B$ is

$$T_{B,II} = T_{B,I} - \Delta E \,, \tag{6.6}$$

where as this becomes

$$T^*_{B,II} = T_{B,I} - \Delta E - E^* \,, \tag{6.7}$$

for the intermediate reaction, thus the difference in beam energy between atoms neutralised with these reactions is given by the excitation energy for the intermediate state,

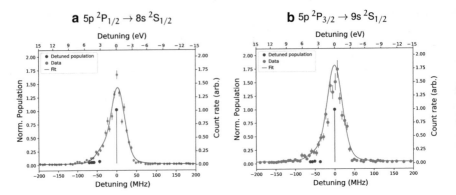

Fig. 6.11 The calculated detuned populations compared to direct population (at 0 MHz) of the lower levels **a** 5p $^2P_{1/2}$ and **b** 5p $^2P_{3/2}$ for the 5p $^2P_{1/2} \rightarrow$ 8s $^2S_{1/2}$ (246.0 nm) and 5p $^2P_{3/2} \rightarrow$ 9s $^2S_{1/2}$ (246.8 nm) transitions respectively. The line profiles for each transition were modified according to these populations. The fitted line profiles also indicated

$$\Delta T^*_{B,ll} = E^* . \tag{6.8}$$

The intermediate states de-excite by spontaneous photon emission [18], the population in the lower state of interest as well as the sidepeak amplitude and asymmetry is therefore a function of time from the neutralisation point. In addition to this source of the sidepeak amplitude, inelastic collisions following neutralisation can have a similar effect [19] (the degree of similarity depends upon the energy spacing of the atomic levels of the atomic system), this contribution is discussed later.

These asymmetries were already simulated as the population passing through the intermediate states was also saved in the simulation. The Voigt line profile was used to model the resonances, with the sidepeaks offset by their corresponding frequency to model the indirect populations via the intermediate states.

In Fig. 6.11 both the simulated population and fitted peak shape for the 5p $^2P_{1/2}$ $\rightarrow$ 8s $^2S_{1/2}$ (246.0 nm) transition is shown, where the largest contribution was determined to be from the $5s^26s$ $^2S_{1/2}$ intermediate state

$$In^+(5s^2\,^1S) + Na(^2S_{1/2}) \rightarrow In(5s^26s\,^2S_{1/2}) + Na^+(^1S) + \Delta E , \tag{6.9}$$

which then de-excites to the ground state

$$In(5s^26s\,^2S_{1/2}) \rightarrow In(5s^25p^2P_{1/2}) + E^*_\gamma , \tag{6.10}$$

for those atomic states $E^*_\gamma = 3.021$ eV of additional kinetic energy required from the ion beam for this reaction, compared to those atoms which directly populate the $5s^26s\,^2S_{1/2}$ state. This difference corresponds to a resonance peak Doppler shifted by 40 MHz to lower frequencies (for the collinear geometry). A much smaller contribution was predicted from charge exchange alone for the 5p $^2P_{3/2} \to$ 9s $^2S_{1/2}$ (246.8 nm) transition, shown in Fig. 6.11b.

Many of the individual contributions from the detuned populations could not be resolved with the laser used (34 MHz linewidth [14]), therefore the amplitude of the largest contribution was used as the free parameter the peak shape fitting, with the others fixed proportional with the simulated relative populations.

The sidepeak amplitudes for the 5p $^2P_{3/2} \to$ 9s $^2S_{1/2}$ (246.8 nm) and 5p $^2P_{1/2}$ $\to$ 8s $^2S_{1/2}$ (246.0 nm) transitions were determined to be 5.7(27)% and 13.9(14)% respectively. These values are an average of 12 hyperfine spectra for the 5p $^2P_{3/2} \to$ 9s $^2S_{1/2}$ (246.8 nm) transition and 10 for 5p $^2P_{1/2} \to$ 8s $^2S_{1/2}$ (246.0 nm) transition. The contributions predicted by simulation were 0.5% and 3.8%, the largest contribution was from the 6s $^2S_{1/2}$ state for both transitions.

The peak asymmetry in hyperfine structure measurements is also attributed to inelastic collisions. Successive collisional excitation and de-excitations into an accessible state with an excitation energy of ΔE creates sidepeaks evenly spaced as $n\Delta E$ for n successive collisions [19]. The probability of n independent successive collisions is given by Poisson's law as

$$P(n) = \frac{x^n}{n!} \exp(-x) \,, \tag{6.11}$$

where x is the factor depending on the length and density of the vapour. The amplitude ratio between the sidepeaks is then given by $P(n)$.

The first excited state after the $^2P_{1/2}$ and $^2P_{3/2}$ states is also the $5s^2 6s^2$ $S_{1/2}$ state. The remaining contribution to the sidepeak amplitude can then be attributed to this mechanism, the remaining contributions being 10.1(14)% and 5.2(27)% respectively. The sidepeak contributions are summarised in Table 6.2. An additional contribution to the sidepeak amplitude which could not be quantified is the population of unknown states during charge exchange which also decay to populate the $^2P_{1/2}$ and $^2P_{3/2}$ states.

An alternative contribution to the side peak amplitudes could be due to the population of states during charge exchange, which decay to the $^2P_{1/2}$ and $^2P_{3/2}$, but which are unknown in literature. The most likely origin of the sidepeaks for symmetric charge exchange (e.g.. Na^+ + Na) is then inelastic collisions (the ground state population will be dominant for low T_B). In the case of asymmetric charge exchange, the higher probability for populating intermediate states will introduce these intrinsic sidepeaks in addition to those from inelastic collisions.

The sidepeak contribution from this source would have been much higher using potassium to neutralise indium, as potassium has a lower ionization potential than sodium and will therefore have a larger relative population in high-lying intermediate states. This can be seen in Fig. 2.8.

Table 6.2 Comparison of simulated sidepeak amplitude contributions from excited upper states to values determined experimentally from fitting the line profile of ^{115}In reference measurements. The contributions were fitted from the lower state of the $^2P_{1/2} \rightarrow 8s\,^2S_{1/2}$ (246.0 nm) and $5p\,^2P_{3/2} \rightarrow 9s\,^2S_{1/2}$ (246.8 nm) transitions. Spectra with $0.8 < \chi_r < 1.2$ were included in the average. 'Collisional' referees to energy loss of the indium atomic beam by collisions with sodium, discussed in the text

	Upper state			
	$6s\,^2S_{1/2}$	$5d\,^2D_{3/2}$	$7s\,^2S_{1/2}$	$6d\,^2D_{3/2}$
$5p\,^2P_{1/2}$ contrib.	2.70%	0.60%	0.30%	0.20%
E_γ (eV)	3.022	4.078	4.501	4.841
Sim. total	3.80%			
Measured	13.9(14)%			
Collisional	10.1(14)%			
$5p\,^2P_{3/2}$ contrib.		0.40%		0.10%
E_γ (eV)		3.807		4.573
Sim. total	0.50%			
Measured	5.7(27)%			
Collisional	5.2(27)%			

6.3   Extracted Atomic Observables

6.3.1   Averaging Observables

The hyperfine structure constants extracted from the scans of the reference isotope ^{115}In are shown for the $^2P_{1/2} \rightarrow 8s\,^2S_{1/2}$ (246.0 nm) transition in Fig. 6.14 and for the $^2P_{3/2} \rightarrow 9s\,^2S_{1/2}$ (246.8 nm) transition in Fig. 6.13. The averaging and uncertainty calculation of the hyperfine structure constants is described below.

The parameters were averaged using the weighted mean $\bar{x}$ as [20]

$$\bar{x} = \frac{\sum_{i=1}^{n} x_i \sigma_i^{-2}}{\sum_{i=1}^{n} \sigma_i^{-2}}, \tag{6.12}$$

where x_i is a single measurement and σ_i is the uncertainty on the measurement. The values are weighted by the reciprocal of the variance $w_i = \frac{1}{\sigma_i^{-2}}$. Initially there was an over dispersion (scatter outside of statistical uncertainty) of the parameter from the individual reference scans due to unknown sources of experimental error not accounted, for this reason the dispersion relation

$$\hat{\sigma}_{\bar{x}}^2 = \frac{1}{\sum_{i=1}^{n} \sigma_i^{-2}} \times \frac{1}{(n-1)} \sum_{i=1}^{n} \frac{(x_i - \bar{x})^2}{\sigma_i^2}, \tag{6.13}$$

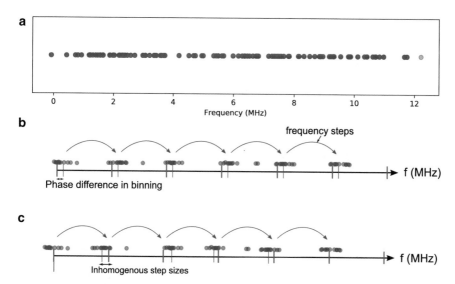

Fig. 6.12 **a** Recorded frequencies of the scanned resonance laser using setpoints of 6 MHz per step. Below are exaggerations of the data (not real) to show possible variations in binned count rate due to **b** a 'phase' difference between the binned and actual laser frequencies scanned and **c** inhomogeneous laser frequency steps. The blue lines indicate the laser frequency step sizes, with the blue dots indicating the spread around the frequency set points, the red line indicates homogeneous frequency bin widths

was used to estimate the uncertainty on the mean value accounting for dispersion, $\hat{\sigma}_{\bar{x}}^2$. Where $\hat{\sigma}_{\bar{x}}^2$ is related to the usual uncertainty on the weighted mean $\sigma_{\bar{x}}^2$ by $\hat{\sigma}_{\bar{x}}^2 = \sigma_{\bar{x}}^2 \chi_r^2$, where the reduced chi-squared χ_r^2 is included to account for any over dispersion [20].

6.3.2 Systematic Error

The statistical error alone of each scan was far too small to account for the dispersion of the hfs constants and isotope shifts, as shown for example for the [115]In reference isotope values in Fig. 6.13. Therefore additional possible sources of systematic error were assessed using the [115]In reference isotopes and included in quadrature to the error of each parameter extracted from the scans of each nuclear state. See Table 6.3 for the systematic contributions which were determined for the final uncertainties.

The assessed sources of additional systematic uncertainty, remaining from the correlations already discussed, included the following:

1. A possible drift in the wavelength accuracy in the range of the quoted accuracy of 2 MHz for the wavemeter (Sect. 4.3). This was assessed by the change in atomic parameters extracted by simulating linear drift in the wave meter of 0 - 2 MHz

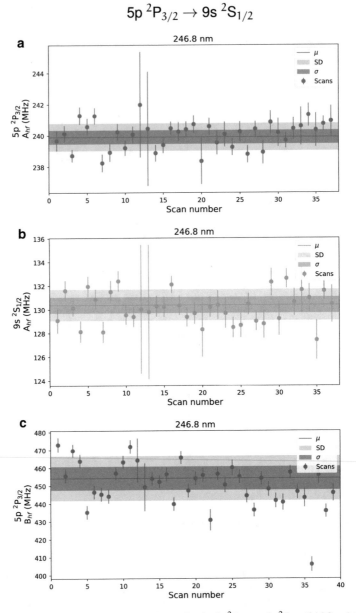

Fig. 6.13 Variation in **a** A_{lower}, **b** A_{upper}, **c** B_{lower} for the $5p\,^2P_{3/2} \rightarrow 9s\,^2S_{1/2}$ (246.8 nm) transition over all scans taken of the reference isotope ^{115}In. μ—Weighted average mean. σ—Weighted average error. SD—Standard deviation

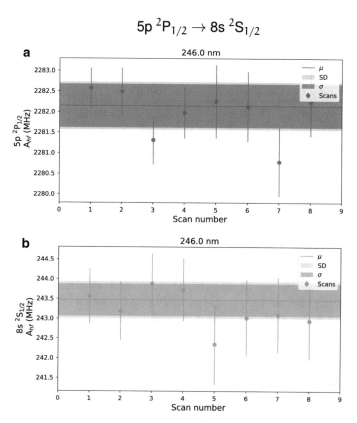

Fig. 6.14 Variation in A) A_{lower}, B) A_{upper} for the 5p ^{2}P$_{1/2}$ → 8s ^{2}S$_{1/2}$ (246.0 nm) transition over all scans taken of the reference isotope ^{115}In. μ—Weighted average mean. σ—Weighted average error. SD—Standard deviation

over the time to take a single scan, using count-rate data from a hyperfine structure scan of ^{115}In.

2. Variation in the extracted parameters with changes in binning of the spectra, in a range of 1 MHz to 20 MHz bins. This was determined from scatter in the fit parameters when re-binning the spectra with bin widths of 1, 3, 6, 12, 16 and 20 MHz.

3. Possible Doppler shift drift due to the limited precision of the extraction voltage readout and short time-scale variations in the extraction voltage. This was determined by comparing the atomic parameters extracted with and without using the ISCOOL Doppler-correction data.

Of these sources the binning had by far the largest effect, this is possibly because of alterations in the line shapes with binning due to fluctuations in the ionization process over the course of a scan. While this source of uncertainty is partially accounted for in the χ_r^2 of the hyperfine structure fit, the extent depends upon the uncertainty in

Table 6.3 Systematic error calculated from varying parameters of a reference scan from each transition. λ cali.—wavemeter calibration uncertainty. Data Analysis—uncertainty from variation in binning and fitting. Voltage cali.—ISCOOL extraction voltage calibration precision. Additional possible sources of remaining systematic uncertainty are discussed in this section

	A_{hf}	A_{hf}	B_{hf}	A_{hf}	A_{hf}
	$5p\,^2P_{3/2}$	$9s\,^2S_{1/2}$	$5p\,^2P_{3/2}$	$5p\,^2P_{1/2}$	$8s\,^2S_{1/2}$
	(MHz)	(MHz)	(MHz)	(MHz)	(MHz)
1. Wavemeter drift	0.02	0.06	0.10	0.03	0.04
2. Frequency binning	0.13	0.37	0.39	0.14	0.07
3. ISCOOL V drift	0.02	0.05	0.14	0.05	0.04
Total	**0.13**	**0.38**	**0.43**	**0.15**	**0.09**

count rate for each bin and therefore the width in frequency of the bins. In reality the laser frequency scan ranges were not perfectly divisible by the laser frequency steps, therefore a beating effect will appear in the count rate as a function of bin width. Additionally, the laser frequency steps were not always precise, but with an oscillation around the set point of the frequency ($\sim$1 MHz). An additional source of variation in the count rate with frequency binning width could be due to inhomogeneity in the actual laser frequency step sizes taken, which could be at the level of the wavemeter relative precision (0.5 MHz), or greater following the correction for the wavemeter frequency drift. Both of these effects could also contribute in isolation or together to explain the dependence of the extracted parameters on the frequency binning used. They are illustrated in Fig. 6.12a–c. An adaptive function to work out the correct minimum inhomogeneous bin sizes for each spectrum could be used in future work to reduce these two effects.

The inclusion of these systematic errors resulted in improved agreement with the average error for the A_{hf} parameters shown in Figs. 6.14 and 6.13. These sources of systematic error were then also included for all the extracted indium nuclear state parameters. A remaining source of error for the parameters remained which could not be quantified however. This is visible in the over dispersion seen in Fig. 6.13a–c. The errors on the ^{115}In parameters were calculated using the over-dispersion correction, Eq. 6.13, this can be seen by the wide error band for the average B_{hf} value.

In order to take into account these sources of systematic uncertainty in the hfs constants extracted for the other indium isotopes, which have comparatively few scans, these uncertainties determined using the ^{115}In isotopes were used. As the source of the over dispersion in the hfs parameters could not be found for the ^{115}In values, the standard deviation measured for the ^{115}In was included as an additional systematic error in the determination of the B_{hf} values for the $^2P_{3/2}$ state.

The source of this dispersion may be due to residual short time-scale variations in the wavemeter drift correction, which are clearly visible at around 75 h in Fig. 6.5a, on-top of the longer time scale variations. This could be verified by applying a sine function correction to the spectra with varying frequency, and comparing the effect to the frequencies of the variation observed in diode reference laser frequency by the wavemeter. However, using the above approach the magnitude of this error was already quantified and accounted using the ^{115}In reference measurements.

6.3.2.1 Isolated or Partial Measurements

Of the spectra collected it is worth highlighting those measurements which may have poorer reliability, as this needed to be accounted for in their reported accuracy. Table 6.1 summarises the completed scans with the two transitions.

As the experiment was time limited, for some lower-priority isotope measurements only single scans of their hyperfine structure were performed. The isotopes ^{117}In, ^{119}In, ^{121}In, ^{123}In, ^{124}In were measured only once with the 5p ^{2}P$_{3/2} \rightarrow$ 9s ^{2}S$_{1/2}$ (246.8 nm) transition. While the isotopes ^{113}In, ^{114}In, ^{116}In, ^{117}In, ^{118}In, ^{122}In, ^{128}In were measured only once with the 5p ^{2}P$_{1/2} \rightarrow$ 8s ^{2}S$_{1/2}$ (246.0 nm) transition. The hyperfine constants determined for these isotopes therefore have larger reported uncertainties as a weighted averaging does not occur, the systematic uncertainties determined in Sect. 6.3.2 dominate.

As some of these single scans were measured once with both transitions, they are later compared for consistency in this analysis.

Two of hyperfine structure scans can be considered as incomplete. ^{122}In measured with 5p ^{2}P$_{1/2} \rightarrow$ 8s ^{2}S$_{1/2}$ (246.0 nm) transition was created from the combination of two scans due to a laser lock error part way through the initial scan, this took 17 min to resolve and continue the scan. While the scan of ^{124}In with the 5p ^{2}P$_{3/2} \rightarrow$ 9s ^{2}S$_{1/2}$ (246.8 nm) transition missed two hyperfine transition peaks. This is visible in the spectra compilation Fig. 6.1 and was simply due to human error in calculating the scan ranges. The values for ^{124}In are therefore not reported in the tables of this work, as it was not possible to quantify the error of the hyperfine parameters without the complete spectra. The ^{116}In measurement with 246.0 nm transition was also a partial scan, with a single peak missing from the 5$^+$ state. The ^{116}In isotope was measured multiple times with the 246.8 nm transition however, therefore in the fitting procedure a linked model was used to fit the states of ^{116}In with both transitions simultaneously, the ratio of A_{hf} between the transitions was taken from ^{115}In.

6.3.2.2 Linked Model Fitting

The atomic parameters fitted in the indium spectra were left as unconstrained as possible to avoid biasing final values. However in some cases a spectrum did not allow for unconstrained fitting. Simultaneous fitting of the spectra from the 5p ^{2}P$_{1/2}$ $\rightarrow$ 8s ^{2}S$_{1/2}$ (246.0 nm) and 5p ^{2}P$_{3/2} \rightarrow$ 9s ^{2}S$_{1/2}$ (246.8 nm) transitions was required

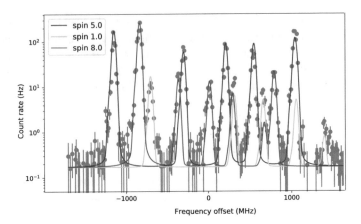

Fig. 6.15 Example ^{116}In spectrum measured with the 5p ^{2}P$_{3/2}$ → 9s ^{2}S$_{1/2}$ (246.8 nm) transition, showing the obscured $I^\pi = 1^+$ state which required constricting of the $I^\pi = 5^+$ and $I^\pi = 8^-$ states with the 5p ^{2}P$_{1/2}$ → 8s ^{2}S$_{1/2}$ (246.0 nm) transition to fit more accurately

to extract the hyperfine parameters of the $I^\pi = 1^+$ state of ^{116}In. This was necessary as many of the hyperfine structure peaks of the $I^\pi = 1^+$ state were obscured by the neighbouring nuclear states for the 246.8 nm transition, while $I^\pi = 1^+$ state was only partially measured with the 246.0 nm transition. Figure 6.15 shows an example spectrum for ^{116}In with the 5p ^{2}P$_{3/2}$ → 9s ^{2}S$_{1/2}$ (246.8 nm) transition which shows the issue. The A_{hf} constant ratios between the two transitions from ^{115}In were used to link the fitting of the spectra. The quadrupole moments extracted from the ^{116}In $I^\pi = 1^+$ state are later shown to be consistent with literature.

In the ^{131}In hyperfine structure spectrum the $I^\pi = 1/2^-$ isomer structure becomes collapsed and unresolvable with the 5p ^{2}P$_{3/2}$ → 9s ^{2}S$_{1/2}$ (246.8 nm) transition, as shown in Fig. 6.27b. However the structure was still resolvable in the 5p ^{2}P$_{1/2}$ → 8s ^{2}S$_{1/2}$ (246.0 nm) transition (Fig. 6.28), due to its larger A_{hfs} parameter. Therefore the linked model fitting was used between the spectra of ^{131}In with the two transitions, essentially constraining the shape of the asymmetry of the collapsed structure seen with the 246.8 nm transition, allowing for a higher accuracy of the centroid value of the $I^\pi = 1/2^-$ structure for the 5p ^{2}P$_{3/2}$ → 9s ^{2}S$_{1/2}$ (246.8 nm) transition.

6.3.3 Reaching the Limit of Sensitivity: ^{132}In

An order of magnitude reduction in yield was expected from ^{130}In to ^{132}In [21]. Only 0.1 pA of ^{130}In was measured on the ISOLDE central beamline Faraday cups with RILIS laser enhancement (2×10^4 ions/s were measured at the ISOLDE table station), out of 2 pA of ^{130}Cs (the ratio of ^{130}In to ^{130}Cs measured at the ISOLDE tape station was 0.1%, from the decay of all nuclear states) and other contamination.

Whereas a beam current of 6 pA was measured at mass 132 with no discernible enhancement from RILIS lasers. Almost all of this can be attributed to ^{132}Cs, which was measured to be produced at a rate of 5×10^8 per μC of proton current, by β^- spectroscopy on the ISOLDE tape station [22].

A beam current of 2 pA was measured at mass 113 with RILIS lasers on and 0.1 pA with RILIS lasers blocked. With this almost pure ^{113}In beam the count rate on resonance was measured to be 1200 cps, indicating a total measurement efficiency from the ISOLDE central beamline to detection by CRIS of 0.05% or, 1:2000. While the maximum count rate for the ^{130}In measurement was 10 cps, corresponding to a total efficiency of 0.1% or 1:1000 using the tape station yields. The difference could be due to saturation of the data-acquisition system at the high ^{113}In count rate, or a difference in transmission compared to the ^{130}In measurement at the tape station.

At mass 132 with 1 pA of beam current (following gating), a count rate of 0.89(17) cps was observed with no lasers going into the CRIS beamline, no increase in count rate outside of this error was seen with lasers on and overlapping the bunch. This implies a small proportion of photo-ionization background from the lasers and a background suppression of 1 in 10^8 of the contamination. No peaks were observed above background in the expected scan range for ^{132}In. Assuming an order of magnitude reduction in yield [23] and a 0.05% total efficiency to give a conservative estimate, a simulated spectrum for ^{132}In is shown in Fig. 6.16. A 3σ effect above background is predicted following 48 h of scanning a 6 GHz range, using coarse frequency steps of 20 MHz. This assumes that the background rate is constant, but this varied as a function of proton current on the UC$_X$ target, a stable proton super-cycle configuration would therefore also be required.

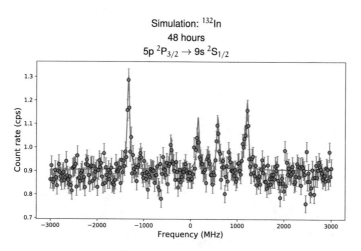

Fig. 6.16 Simulated spectrum of ^{132}In using the background and signal rates in the text. For the $5p\,^2P_{3/2} \rightarrow 9s\,^2S_{1/2}$ (246.8 nm) transition, assuming a spin of $I = 7$ for the ground state of ^{132}In. Using 9.6 min per 20 MHz frequency step

Future measurements of neutron rich-indium at ISOLDE with the CRIS experiment would require a large reduction in this background to continue to measure lower yield neutron-rich isotopes without further significant increases in measurement time, although the UC$_x$ target was accidentally vented during this experiment which increased the level of caesium contamination. As the 1064 nm light had little influence on the background count rate, it can be assumed that non-resonant photo ionization of the contaminant component was not the dominant source of background from the contamination.

Collisional ionization of caesium was the most likely source of background, as it was the most likely contamination in the mass region of interest (115–132). Isobaric contamination can be orders of magnitude larger than the isotope of interest at ISOL facilities [24], allowing collisional ionization to result in an appreciable non-resonant background. The pressure of the interaction region was $1-2\times10^{-9}$ mbar with all valves open during the experiment.

During the charge-exchange process many high-lying states are also populated, by the element of interest as well as the contamination. This can be seen in the relative population simulation Fig. 2.8 for indium, in Sect. 2.4. Many high-lying states (including Rydberg states) can have long lifetimes [25], however at these much higher atomic energy levels the geometric cross section for collisional ionization becomes much larger [26, 27]. This source of background can partly be quantified by looking at the fraction of atomic population in these high-lying states following charge exchange, using the simulations outlined in Sects. 2.4 and 5.9.

In Fig. 6.17 the fraction of the atomic population in these high-lying states is shown for $1 \leq Z \leq 89$. For the energies below the ionization potential of the element, the Nd:YAG laser harmonics wavelengths are plotted (1064 nm, 532 nm, 355 nm and 266 nm), giving a guide as to the extent these states will contribute to the background for a given atomic number. Although the simulations may overestimate these popula-

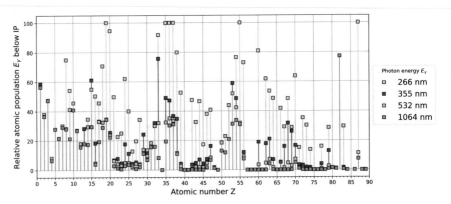

Fig. 6.17 Calculated fraction of atomic population within photon energies E_γ of the ionization potential of elements $1 \leq Z \leq 89$. Using the simulations discussed in Sects. 2.4 and 5.9. Cross sections were calculated for a 40-keV ion beam neutralised by a sodium vapour. Populations are stated after 120 cm of flight

tions if transition rates are not fully known, the relative population in the plot indicates the level of certainty one can have these high-lying states will have a contribution to laser or collisional background for a given element.

For caesium ($Z = 55$), ~30% of its atomic population is expected to be within reach of the energy of the ionization potential for being ionized by 1064 nm light, used as the final non-resonant ionization step laser for the neutron-rich indium isotope measurements. As no increase in background was observed with the 1064 nm light overlapping the atom bunch in the interaction region, it is possible that the photo-ionization cross section for these high-lying states may be low [28] (no weighting was included for the photo-ionization cross sections from the states) or that the contribution was collisionally re-ionized before interacting with the 1064 nm light in the poorer vacuum region along the charge-exchange cell ($<10^{-6}$ mbar).

This source of collisional background can be reduced by the use of field ionization plates [29, 30] after the charge-exchange cell to ionize and deflect the atoms in these high-lying states before they reach the interaction region. These field ionization plates were recently implemented and tested in the $^{104-115}$In mass region and indeed resulted in a factor of 2–3 reduction in the background rate for indium [31].

Further reduction of collisional ionization can be achieved by reducing the pressure in the interaction region, improvements have been made since the neutron-rich indium experiment measurements, allowing a reduction down to 3×10^{-10} mbar [32]. This may be difficult to further reduce without the use of cryogenic cooling of the beamline.

It is also likely that the collisional background is increased due to a molecular beaming effect (mentioned in Sect. 3.2.1) of neutrals which escape the relatively poor vacuum region of the charge-exchange cell and effectively higher-density region for collisional ionization which overlaps with the path of atom bunch to be probed (see Fig. 3.10).

Alternative detection methods can be used to increase the background suppression. Field ionization with the appropriate excitation scheme to a Rydberg state can be used to ionize only those atoms below a known ionization potential at the end of the beamline [29, 33]. In cases where the α- or β^--decay half-life of the isotope is sufficiently short or the γ-decay transitions are well known then decay-assisted laser spectroscopy can be used to suppress background contamination [34]. A variety of methods on the production, ionization and separation side of ISOLDE can also be used purify the radioactive beams [22, 35], which could be further developed for indium measurements.

6.3.4 Hyperfine Parameters and Isotope Shifts

Taking into account the above considerations in the analysis, the extracted isotope and isomer shift values relative to ^{115}In are displayed in Table 6.5. Although the error is larger for some of the lower-yield isomers, the isomer shifts are clearly determined and their implications will be discussed later.

The extracted hyperfine parameters are displayed in Table 6.4, for two of these atomic states, 5p $^2P_{1/2}$ and 5p $^2P_{3/2}$, previous measurements by [36] were included. It is informative to compare the values at this stage, before additional error may be introduced in converting to the nuclear electromagnetic moments.

The ^{113}In and ^{115}In isotope A_{hf} and B_{hf} values shown in Table 6.4 were all found to agree within 1σ of the values determined using the ablation ion source, displayed in Table 5.7 of Chap. 5. The isotope shift values determined between ^{113}In and ^{115}In in Table 5.5 using the ablation source also have 1σ agreement with literature, which gives further validation of the time-of-flight correction technique used to determine the isotope shifts (described in Chap. 5).

All of the newly determined hyperfine parameters are within 2σ of literature values [36], with the following exceptions: ^{116}In 5$^+$ for the A_{hf} of the 5p $^2P_{1/2}$ state, ^{117}In 1/2 $^-$ for the A_{hf} of the 5p $^2P_{3/2}$ state, which are within 5σ, and 4σ respectively. Of the extracted B_{hf} values, most agree within 1σ of the values reported by [36], the B_{hf} values for the ^{118}In 5$^+$ and ^{116}In 5$^+$, ^{119}In 9/2 $^+$ and ^{120}In 8$^-$ states agree by 2σ.

The discrepancies could be due to an additional uncertainty not accounted for in the fitting of these spectra, which have a few overlapping hyperfine transition peaks, or intermittent undetermined experimental error, those within 2σ are reasonable within scatter of the large data set. Only single scans were performed at each of these masses (^{116,117,118}In). Additionally, the literature values reported by [36] did not report any systematic uncertainties. Comparison with the μ and Q_S values determined from these parameters will provide a test of the agreement of the parameters determined with each of the atomic states, and also indicate the physical significance of the values outside of 2σ.

6.4   Extraction of Nuclear Moments

6.4.1   Magnetic Dipole Moments

The magnetic moments were extracted using the hyperfine constant ratio (Eq. 2.29) with the magnetic moment of ^{115}In as the reference, the value was $\mu = +5.5408(2)\,\mu_N$ as measured by a high-precision NMR experiment [38]. Which was found to agree with the value of $\mu = +5.560(62)\,\mu_N$ determined with the calculations and our measured A_{hf} value for ^{115}In, displayed in Table 5.8.

The extracted parameters are displayed in Table 6.6, where four columns display the magnetic moments from the four atomic states from which the hyperfine constants were measured. The literature values of the magnetic moments are also displayed, a variety of techniques were used to determine these values. Most measurements heavier than ^{118}In were measured by [36], also using laser spectroscopy.

Table 6.4 Compilation of nuclear spins and hyperfine structure constants extracted from this work using the transitions at 246.0 nm (5p $^2P_{1/2} \rightarrow$ 8s $^2S_{1/2}$) and 246.8 nm (5p $^2P_{3/2} \rightarrow$ 9s $^2S_{1/2}$). All values have positive signs except where sign is given. Literature values were available from [36], with the exception of the values denoted by †, which are reported by [37]

Mass	Spin	This work					Literature		
		A_{hf} 5p $^2P_{3/2}$ (MHz)	A_{hf} 9s $^2S_{1/2}$ (MHz)	B_{hf} 5p $^2P_{3/2}$ (MHz)	A_{hf} 5p $^2P_{1/2}$ (MHz)	A_{hf} 8s $^2S_{1/2}$ (MHz)	A_{hf} 5p $^2P_{1/2}$ (MHz)	A_{hf} 5p $^2P_{3/2}$ (MHz)	B_{hf} 5p $^2P_{3/2}$ (MHz)
113	$9/2^+$	242.44(94)	129.3(14)	441(15)	2276.0(8)	242.66(84)	2277.086	241.6409(4) †	443.414(4)
113	$1/2^-$	−82.6(65)	−50(19)		−774(49)	−90(53)			
114	1^+	556.4(57)	301(14)	57(15)					
114	5^+	182.76(91)	99.6(14)	412(12)	1729.24(66)	184.96(44)	1729.0(2.0)	183.5(0.4)	410(6)
115	$9/2^+$	239.9(16)	130.3(14)	452(24)	2281.93(84)	243.29(59)	2281.9501(4)	242.1647(3) †	449.545(3)
115	$1/2^-$	−94.5(21)	−47.3(79)		−903(20)	−66(11)	−902.5(1.1)	−95.97(1)	
116	1^+	554.7(84)	292(19)	46(14)					
116	5^+	166.54(94)	92.7(14)	421(13)	1566.6(13)	169.3(21)	1573.7(0.4)	168.3(0.5)	445(6)
116	8^-	79.05(93)	43.6(14)	158(13)	743.63(77)	79(1)	744.6(0.4)	80.1(0.3)	172(5)
117	$9/2^+$	240.65(92)	129.7(14)	465(12)	2277.3(18)	243.5(15)	2279.2(1.0)	241.4(0.5)	460(6)
117	$1/2^-$	−106.9(28)	−51(10)		−1028(11)	−112.8(78)	−932.996(12)	−99.005(10)	
118	1^+	546.6(25)	292.4(61)	71(13)	5142.8(19)	548.5(35)			
118	$(5)^+$	166.21(93)	89.9(14)	425(13)	1567.48(74)	167.44(88)	1568.9(1.6)	167.1(0.3)	442(4)
118	(8^-)	81.91(91)	44.7(13)	243(13)	768.35(69)	82.39(46)	770.1(0.4)	82.5(0.2)	245(5)
119	$9/2^+$	240.39(93)	130.5(14)	457(13)			2270.6(1.2)	241.7(0.3)	474(4)
119	$1/2^-$	−134.9(47)	−70(10)					−124.8(4.3)	
120	1^+	547.0(48)	293(11)	82(13)	5133.6(19)	543.4(63)			
120	5^+	170.38(97)	94.4(15)	439(13)	1590.8(9)	170.64(66)	1591.7(0.8)	168.9(0.8)	450(9)
120	8^-	91.22(91)	49.6(13)	277(12)	855.45(56)	91.13(44)	854.9(0.5)	91(2)	294(5)
121	$9/2^+$	242.93(91)	132.8(13)	462(13)			2266.0(1.8)	240.8(0.3)	452(6)
121	$1/2^-$	−140.3(17)	−87.8(33)		5140.3(29)			−139.0(3.5)	
122	1^+	546.4(51)	293(12)	109(12)		548.3(64)			
122	5^+	172(1)	94.3(17)	452(13)	1599.2(11)	171.02(54)	1600.1(1.3)	170.1(0.9)	447(11)
122	(8^-)	93.20(91)	50.3(13)	337(13)	875.22(65)	93.64(45)	875.5(0.7)	93.0(0.3)	329(12)

(continued)

Table 6.4 (continued)

Mass	Spin	This work					Literature		
		A_{hf} 5p $^2P_{3/2}$ (MHz)	A_{hf} 9s $^2S_{1/2}$ (MHz)	B_{hf} 5p $^2P_{3/2}$ (MHz)	A_{hf} 5p $^2P_{1/2}$ (MHz)	A_{hf} 8s $^2S_{1/2}$ (MHz)	A_{hf} 5p $^2P_{1/2}$ (MHz)	A_{hf} 5p $^2P_{3/2}$ (MHz)	B_{hf} 5p $^2P_{3/2}$ (MHz)
123	$9/2^+$	237.68(92)	129.2(14)	424(13)			2264.6 (1.3)	240.4 (0.3)	420 (5)
123	$1/2^-$	−159.9(22)	−80.3(47)					−160.3 (5.2)	
125	$9/2^+$	240.3(11)	129.9(15)	387(14)			2265.9 (1.4)	240.5 (1.0)	394 (20)
125	$1/2^-$	−177.5(71)	−92(15)					−172.0(4.4)	
126	$3^{(+)}$	262.3(11)	141.4(17)	275(13)			2497.5(7.5)	264.0(5.7)	274(32)
126	(8^-)	99.63(93)	55.0(14)	380(13)			943.3(1.7)	100.1 (1.6)	379(7.0)
127	$9/2^+$	242.36(93)	130.1(14)	338(16)	2278.30(58)	243.80(45)		240.8(0.7)	327(16)
127	$1/2^-$	−172.4(27)	−87.0(57)		−1613.3(94)	−174.1(81)			
127	$(21/2^-)$	99.22(93)	54.2(14)	464(12)	942.51(68)	100.89(66)			
128	3^+	269.84(99)	147.9(15)	239(13)	2525.2(12)	272.4(21)			
128	(8^-)	114.78(93)	61.9(14)	266(12)	1078.36(77)	114.72(75)			
129	$9/2^+$	243.5(18)	132.63(72)	280.4(73)	2304.86(95)	244.84(72)			
129	$1/2^-$	−157.7(25)	−81.1(13)		−1434.5(22)	−162(14)			
129	$(23/2^-)$	100.99(96)	54.69(26)	344(18)	956.61(64)	101.30(42)			
130	1^-	−689.9(63)	−381(11)	34.5(55)	−6400.3(63)	−682.7(77)			
130	(5^+)	114.2(16)	61.3(14)	214.2(87)	1063.25(67)	113.19(92)			
130	(10^-)	99.0(5)	53.8(8)	333(15)	929.00(84)	99.42(52)			
131	$9/2^+$	275.9(6)	149.3(7)	177.3(57)					
131	$1/2^-$	−19.7(74)	−11(4)		−188(15)	−20.3(19)			

Table 6.5 Compilation of isotope shifts extracted in this work using the transitions at 246.0 nm (5p $^2P_{1/2} \to$ 8s $^2S_{1/2}$) and 246.8 nm (5p $^2P_{3/2} \to$ 9s $^2S_{1/2}$)

Mass	Spin	$IS^{115,A}$ 246.8 nm (MHz)	$IS^{115,A}$ 246.0 nm (MHz)
113	$9/2^+$	−278(5)	−265(5)
113	$1/2^-$	−238(6)	−241(10)
114	1^+	−188(7)	
114	5^+	−171(10)	−175(5)
115	$9/2^+$	0	0
115	$1/2^-$	33(5)	26(8)
116	1^+	106(20)	
116	5^+	99(20)	89(5)
116	8^-	99(2)	86(8)
117	$9/2^+$	265(3)	243(5)
117	$1/2^-$	283(4)	265(5)
118	1^+	333(3)	334(6)
118	$(5)^+$	329(2)	330(5)
118	(8^-)	324(3)	324(5)
119	$9/2^+$	475(3)	
119	$1/2^-$	487(4)	
120	1^+	571(9)	551(6)
120	5^+	556(5)	531(5)
120	8^-	530(2)	500(5)
121	$9/2^+$	654(2)	
121	$1/2^-$	661(3)	
122	1^+	710(8)	728(7)
122	5^+	674(5)	704(5)
122	(8^-)	658(8)	687(5)
123	$9/2^+$	756(3)	
123	$1/2^-$	753(3)	
125	$9/2^+$	941(4)	
125	$1/2^-$	926(5)	
126	$3^{(+)}$	1026(3)	
126	(8^-)	1019(5)	
127	$9/2^+$	1129(4)	1115(5)
127	$1/2^-$	1114(3)	1098(5)
127	$(21/2^-)$	1110(3)	1086(5)
128	3^+	1147(3)	1147(9)
128	(8^-)	1127(4)	1101(7)
129	$9/2^+$	1251(2)	1254(5)
129	$1/2^-$	1231(4)	1233(6)
129	$(23/2^-)$	1171(2)	1167(5)
130	1^-	1340(3)	1324(8)
130	(5^+)	1281(3)	1253(6)
130	(10^-)	1305(3)	1289(5)
131	$9/2^+$	1364(4)	
131	$1/2^-$	1366(3)	1342(9)

The extracted magnetic moments were found to be consistent within 1 or 2 σ between atomic states. The largest discrepancy is for the $1/2^-$ isomer state of ^{115}In, measured with the 8s $^2S_{1/2}$ state, with a value of $-0.169(29)$ μ_N compared to the other states within error of the NMR measurement of $-0.24398(5)$ μ_N [39]. This may be due to the smaller value of the A_{hf} with only two peaks for this state relative to the 5p $^2P_{1/2}$ state ($A_{hf} = 243.3(6)$ MHz vs 2281.9501(4) MHz) of the same transition (see Fig. 6.27), making it more sensitive to experimental fluctuations. Due to the 100 min half-life of the $1/2^-$ state relative to the stable $9/2^+$ state, a large difference in yield was observed during reference measurements of ^{115}In, which was worsened by turning the protons off during reference measurements (to conserve the UC_x target). Given the large number of measurements performed over the experiment the 2σ deviation for the $1/2^-$ state is likely then not significant. The 5p $^2P_{1/2}$ state provides the most accurate value for the $I^\pi = 1/2^-$ state in ^{115}In. As the hfs constants determined from each of these atomic states constitutes a separate measurement, and the values agree within error, the error-weighted average of the states was taken and is shown in the column $\bar{\mu}$ in Table 6.6. Most of the $\bar{\mu}$ values agree within 2σ of the literature values, and improve upon the precision in some cases. Discrepancies exist between the $\bar{\mu}$ values and the literature values for the ^{121,119}In $9/2^+$ states and the ^{119,117}In $1/2^-$. However in these cases a 1σ agreement between the values from all the atomic states of the separate transitions was found, which gives confidence in reporting these values.

6.4.2 Electric Quadrupole Moments

It is also possible use the hyperfine constants ratio in Eq. 2.30 with the reference isotope to determine the quadrupole moment of the newly measured isotopes, by using literature values. However the quadrupole moments have not yet been measured directly and depend on calculated atomic parameters used in a particular study. Earlier in this work measurements were made of ^{115}In and ^{113}In to benchmark against calculations of the atomic parameters needed to determine the magnetic dipole moment. These same calculations gave a value of the electric field gradient which should be the most accurate available for indium to date and is therefore used for the extraction of the quadrupole moments of these newly measured isotopes. A value of $B_{hf}/Q = 576(4)$ (MHz/b) was calculated for the atomic parameter used to extract the quadrupole moments from B_{hf}. The agreement of the calculated A_{hf}/g_I factors with experimental values, discussed in Sect. 5.6, gave confidence in the calculated B_{hf}/Q factor, which was used to determine the Q_S values listed in Table 6.6.

Many of available literature values in this case were again measured by [36] from the same atomic state 5p $^2P_{3/2}$. The quadrupole moments values are given as originally reported, and recalculated from their B_{hf} values using the new B_{hf}/Q value of this work for consistency. The original and recalculated values are shown in Table 6.6 as Q_S^{lit} and $Q_S^{lit\dagger}$ respectively, as expected the overall agreement improves when using the same B_{hf}/Q factor. Between these literature values and those extracted in this

(5p ^{2}P$_{1/2}$ → 8s ^{2}S$_{1/2}$) and 246.8 nm (5p ^{2}P$_{3/2}$ → 9s ^{2}S$_{1/2}$). *—Literature values reported by [36], all other literature values agree with these quadrupoles, where they do not they also disagree with our measurements and are discussed in the text. †—These values were determined from literature B_{hf} values [36] using the B_{hf}/Q constant calculated in this work

Mass	Spin	5p ^{2}P$_{3/2}$ μ (μ_N)	9s ^{2}S$_{1/2}$ μ (μ_N)	5p ^{2}P$_{1/2}$ μ (μ_N)	8s ^{2}S$_{1/2}$ μ (μ_N)	$\bar{\mu}$ (μ_N)	μ^{lit} (μ_N)	Ref	Q_S (b)	5p ^{2}P$_{3/2}$ Q_S^{lit} (b)	$Q_S^{lit\dagger}$ (b)
113	9/2+	5.547(22)	5.45(9)	5.526(2)	5.526(23)	5.5264(19)	5.5289(2)	[40]	0.767(27)	0.80(4)	0.77(4)
113	1/2−	−0.210(17)	−0.233(88)	−0.209(13)	−0.23(13)	−0.21(1)	−0.21074(2)	[41]			
114	1+	2.829(29)	2.82(14)			2.828(28)	2.817(11)	[42]	0.099(26)		
114	5+	4.646(23)	4.661(86)	4.6653(18)	4.680(16)	4.6654(18)	4.653(5)	[36]	0.716(22)	0.74(12)	0.71(12)
115	9/2+	5.489(36)	5.489(91)	5.541(2)	5.541(19)	5.541(2)	5.5408(2)	[38]	0.784(42)	0.81(5)	
115	1/2−	−0.2402(54)	−0.221(37)	−0.2436(53)	−0.167(29)	−0.2405(38)	−0.24398(5)	[39]			
116	1+	2.820(43)	2.73(18)			2.816(41)	2.7876(6)	[43]	0.079(25)	0.11(1) [44]	
116	5+	4.234(24)	4.338(84)	4.2265(34)	4.283(55)	4.2270(34)	4.235(15)	[36]	0.732(23)	0.802(12)	0.775(12)
116	8−	3.216(38)	3.27(11)	3.2100(33)	3.186(42)	3.2099(33)	3.215(11)	[36]	0.275(23)	0.310(9)	0.300(9)
117	9/2+	5.506(21)	5.465(89)	5.5295(44)	5.546(37)	5.5286(43)	5.519(4)	[36]	0.807(22)	0.83(1)	0.80(1)
117	1/2−	−0.2717(71)	−0.237(48)	−0.2774(29)	−0.29(2)	−0.2766(27)	−0.25174(3)	[36]			
118	1+	2.779(13)	2.737(67)	2.775(1)	2.776(19)	2.775(1)			0.124(22)		
118	(5)+	4.226(24)	4.210(83)	4.229(2)	4.237(25)	4.229(2)	4.231(9)	[36]	0.738(22)	0.796(8)	0.770(8)
118	(8−)	3.332(37)	3.35(11)	3.317(3)	3.34(2)	3.317(3)	3.321(11)	[36]	0.422(22)	0.441(7)	0.426(7)
119	9/2+	5.500(21)	5.497(91)			5.499(62)	5.515(1)	[36]	0.794(23)	0.854(7)	0.826(7)
119	1/2−	−0.343(12)	−0.330(49)			−0.342(12)	−0.319(5)	[36]			
120	1+	2.781(24)	2.74(11)	2.7700(11)	2.750(32)	2.770(1)			0.143(22)		
120	5+	4.331(25)	4.418(89)	4.2918(24)	4.32(2)	4.2926(24)	4.295(5)	[36]	0.763(23)	0.81(2)	0.78(2)
120	8−	3.710(37)	3.72(11)	3.6927(24)	3.69(2)	3.6927(24)	3.692(4)	[36]	0.482(21)	0.530(1)	0.512(1)
121	9/2+	5.558(21)	5.60(9)			5.575(62)	5.502(5)	[36]	0.803(23)	0.814(11)	0.787(11)
121	1/2−	−0.3567(42)	−0.411(16)			−0.3600(41)	−0.355(4)	[36]			
122	1+	2.778(26)	2.74(12)	2.7736(15)	2.775(33)	2.7736(15)			0.19(2)		
122	5+	4.370(26)	4.416(95)	4.3146(29)	4.328(17)	4.3156(28)	4.318(5)	[36]	0.785(24)	0.81(2)	0.78(2)
122	(8−)	3.791(37)	3.76(11)	3.7780(28)	3.79(2)	3.7783(28)	3.781(6)	[36]	0.586(22)	0.59(2)	0.57(2)

(continued)

Table 6.6 (continued)

Mass	Spin	5p ^{2}P$_{3/2}$ μ	9s ^{2}S$_{1/2}$ μ	5p ^{2}P$_{1/2}$ μ	8s ^{2}S$_{1/2}$ μ	$\bar{\mu}$	μ^{lit}	Ref	Q_S	5p ^{2}P$_{3/2}$ Q_S^{lit}	$Q_S^{lit\dagger}$
123	$9/2^+$	5.438(21)	5.445(89)			5.442(61)	5.491(7)	[36]	0.736(23)	0.757(9)	0.732(9)
123	$1/2^-$	−0.4065(56)	−0.376(22)			−0.4047(54)	−0.400(4)	[36]			
125	$9/2^+$	5.498(25)	5.473(94)			5.496(24)	5.502(9)	[36]	0.673(24)	0.71(4)	0.69(4)
125	$1/2^-$	−0.451(18)	−0.430(69)			−0.450(17)	−0.433(4)	[36]			
126	$3^{(+)}$	4.002(17)	3.97(7)			4.000(16)	4.034(11)	[36]	0.477(23)	0.49(5)	0.47(5)
126	(8^-)	4.053(38)	4.12(12)			4.059(36)	4.061(4)	[36]	0.661(23)	0.683(12)	0.660(12)
127	$9/2^+$	5.545(21)	5.48(9)	5.5319(14)	5.552(17)	5.5321(14)	5.522(8)	[36]	0.588(29)	0.59(3)	0.57(3)
127	$1/2^-$	−0.4382(69)	−0.407(27)	−0.4352(25)	−0.44(2)	−0.4355(24)					
127	$(21/2^-)$	5.30(5)	5.33(15)	5.3399(39)	5.362(38)	5.3398(38)			0.806(22)		
128	3^+	4.116(15)	4.154(68)	4.0877(19)	4.135(33)	4.0883(19)			0.416(22)		
128	(8^-)	4.669(38)	4.63(12)	4.6549(33)	4.645(32)	4.6549(33)			0.462(21)		
129	$9/2^+$	5.571(42)	5.588(76)	5.5964(23)	5.576(21)	5.5961(23)			0.487(13)		
129	$1/2^-$	−0.4010(64)	−0.3797(76)	−0.38701(59)	−0.410(34)	−0.38709(58)					
129	$(23/2^-)$	5.905(56)	5.889(79)	5.936(4)	5.896(28)	5.9349(39)			0.598(32)		
130	1^-	−3.508(32)	−3.56(11)	−3.4535(34)	−3.45(4)	−3.4542(34)			0.0599(96)		
130	(5^+)	2.905(42)	2.870(75)	2.8685(18)	2.864(24)	2.8686(18)			0.372(15)		
130	(10^-)	5.034(25)	5.037(98)	5.0127(45)	5.032(29)	5.0138(44)			0.578(26)		
131	$9/2^+$	6.313(14)	6.291(84)			6.312(14)			0.31(1)		
131	$1/2^-$	−0.050(19)	−0.050(19)	−0.0507(41)	−0.0514(48)	−0.051(3)					

work, only those which initially disagreed on their B_{hf} values by 2σ (in Table 6.4) disagree on their Q_S values. As Q_S could only be determined using the 5p ^{2}P$_{3/2}$ state from the 5p ^{2}P$_{3/2} \rightarrow$ 9s ^{2}S$_{1/2}$ (246.8 nm) transition, it was not possible to compare against a value from the other states (5p ^{2}P$_{1/2}$, 8s ^{2}S$_{1/2}$, 9s ^{2}S$_{1/2}$). Further measurements would be required to determine if these discrepancies are significant.

6.5 Extraction of the Nuclear Mean-Square Charge Radii

The well known relation to extract the changes in mean-square charge radii from the measured isotope shift values

$$\delta \left\langle r^2 \right\rangle^{A,A'} = \frac{IS_i^{A,A'} - M_i^{A,A'}}{F_i} \tag{6.14}$$

was used in this thesis for transitions i. Where $M_i^{A,A'} = (K_{NMS,i} + K_{SMS,i})/\mu_{A,A'}$, the mass shift (MS) factor, being the sum of the normal mass shift (NMS) and specific mass shift (SMS). F_i is the field shift (FS) factor. The choice of the atomic parameters K_{SMS} and F_i however requires some discussion.

The tin ($Z = 50$) and cadmium ($Z = 48$) isotopes neighbouring indium have 10 and 8 stable isotopes respectively, allowing absolute $\left\langle r^2 \right\rangle^{A,A'}$ values to be precisely determined from measurements with muonic atoms and elastic electron scattering experiments [45, 46]. This allows extraction of the atomic parameters $M_i^{A,A'}$ and F_i with high precision, needed for obtaining the atomic factors for the evaluation of $\delta \left\langle r^2 \right\rangle^{A,A'}$ from optical spectroscopy measurements for indium. The absolute $\left\langle r^2 \right\rangle^{A,A'}$ values have only been determined using the naturally occurring ^{113,115}In isotopes. This results in a large error of the extracted atomic factors, which makes the extrapolation of $\delta \left\langle r^2 \right\rangle^{A,A'}$ to neuron-rich (or deficient) indium isotopes unreliable. All odd-proton elements, with the exception of potassium, have ≤ 2 naturally abundant isotopes accessible for measurement and therefore share the problem in common.

Two approaches have been used in an attempt to determine the atomic parameters for evaluation of $\delta \left\langle r^2 \right\rangle^{A,A'}$ despite this problem. Both were applied on measurements from the 5p ^{2}P$_{3/2} \rightarrow$ 6s ^{2}S$_{1/2}$ (451 nm) transition [36] in indium and highlight the need for accurate atomic structure calculations.

The first approach by [36] used a relativistic Dirac-Fock calculated value of $F_{451} = 2.07(1)$ GHz/fm^2 [47], combined with an empirically determined $M_i^{A,A'}$ value, this had to be empirically determined due to the difficulty of calculating the $K_{SMS,i}$ contribution to $M_i^{A,A'}$. Charge radii measurements of $\delta \left\langle r^2 \right\rangle$ (^{114,116}Sn) and $\delta \left\langle r^2 \right\rangle$ (^{112,114}Cd) gave a ratio of SMS$^{A,A'} = -1.17(20)$NMS$^{A,A'}$ for the specific to normal mass shift using the analysis described in [48], which allowed the $M_i^{A,A'}$ value to be approximately determined for the transition. This ratio at least suggests that the SMS is a significant effect in the indium atomic system that should be quantified.

The second approach followed muonic atom transition measurements of ^{113,115}In [49] to determine $\delta \langle r^2 \rangle_{o\mu e}^{113,115} = 157(11)$ fm^2. A combined analysis by [50] used the $\delta \langle r^2 \rangle_{o\mu e}^{113,115}$ value combined with the 5p ^{2}P$_{3/2} \rightarrow$ 6s ^{2}S$_{1/2}$ (451 nm) transition optical data to determine $F_{451\ nm} = 1.36$ GHz/fm^2 and $M_{451\ nm} = 364$ GHz. However, the SMS had to be assumed to be SMS$^{A,A'} = 0$ for their analysis and so no error was reported with these atomic factors.

The range of uncertainty between the above two approaches to extract the $\delta \langle r^2 \rangle^{A,A'}$ values from the measured isotope shifts, using the 5p ^{2}P$_{3/2} \rightarrow$ 6s ^{2}S$_{1/2}$ (451 nm) transition [36], is shown in Fig. 6.20a as the orange area 'literature'. The nuclear structure implications of the $^{113-131}$In $\delta \langle r^2 \rangle^{A,A'}$ isotope values is discussed in Chap. 7.

The values extracted using the original data by [36] agree within error of the values calculated using the above parameters by [50], despite the very different atomic parameters used. This is possible as the large negative SMS is cancelled by the larger field shift in the opposite direction. The errors on the F_i factor have more of an effect at higher or lower mass isotopes where the agreement becomes poorer along with the reliability of the extracted $\delta \langle r^2 \rangle^{A,A'}$ values. The large variation in the $\delta \langle r^2 \rangle^{A,A'}$ values and their large error with these two approaches highlights the reliability issues appearing from assuming a SMS factor or inaccurate calculation of the FS factor for the indium atom.

With the new measurements described in this thesis and those in literature [36], there now exists measurement of the isotope shifts in indium over a range of 24 mass units with four separate transitions. Additionally, the measurements from this thesis had sufficient precision for MS-independent isomer shift values to be reported. Using these measurements the FS and the SMS factors could be extracted for comparison to the calculations of the FS and MS factors from the AR-RCC approach mentioned in Sect. 2.3.2. The following section describes the findings.

6.5.1 Comparison to Atomic Factors from Relativistic Coupled Cluster Calculations

Due to the inaccuracy of the approaches mentioned above, calculations of the atomic FS and MS factors were performed in [51] with an 'analytic response' (AR) [52, 53] approach in the relativistic coupled cluster (RCC) theory framework (see Sect. 2.3.2). Two commonly used RCC approaches to determine the MS and FS factors, finite field (FF) and expectation value evaluation (EVE) were also used to calculate the FS and MS factors and the values shown for comparison.

An analysis using the isotope shift measurements from this work enabled all three approaches of the RCC calculations to be compared in order to judge the accuracy of the approaches for calculating the MS and FS factors, and therefore the accuracy of the evaluated $\langle r^2 \rangle^{A,A'}$ values using those factors.

Table 6.7 Comparison of FS, NMS and SMS factors of the six states in indium from the FF, EVE and AR approaches obtained using the RCCSD method. Calculations performed by [55]

Method	$5P_{1/2}$	$5P_{3/2}$	$6S_{1/2}$	$7S_{1/2}$	$8S_{1/2}$	$9S_{1/2}$
F (GHz/fm^2)						
FF	1.544	1.491	−0.437	−0.155	−0.069	−0.033
EVE	1.275	1.299	−0.408	−0.135	−0.061	−0.033
AR	1.435(6)	1.442(6)	−0.383(1)	−0.1281(5)	−0.0559(25)	−0.0307(5)
K_{SMS} (GHz.u)						
FF	749	711	364	170	98	63
EVE	1340	375	458	201	113	71
AR	774(41)	734(37)	340(5)	163(2)	96(1)	61.7(5)
$K_{SMS}^{non-rel.}$ (GHz.u)						
FF	724	583	405	183	105	66
EVE	1355	227	503	216	120	74
AR†	806(57)	670(101)	385(50)	178(17)	103(8)	65(5)
'Energy'	768	731	367	171	99	65
K_{SMS} (GHz.u)						
FF	−470	−403	119	38	17	9
EVE	−1048	−899	136	42	18	10
AR	−638(71)	−533(69)	94(26)	29(8)	13(4)	8.6(5)
Expt.	−536(122)	−507(111)	169(51)	55(42)	24(80)	−13(66)
LIS^{113,115} (MHz)						
Exp.	277(10)	272(6) [48]	17(6)	12(6)	9(12)	2(10)

†Level energies from [56] were used.
* To determine K_{SMS} from Eq. 6.14, the measured differential ISs, $\delta E^{113,115}$, were combined with FS factors from the AR approach and $\delta \langle r^2 \rangle_\mu^{113,115} = 0.157(11)$ fm^2 [57]

The FS, NMS and SMS factors calculated for the atomic states in the 246.8 nm (5p $^2P_{3/2} \to$ 9s $^2S_{1/2}$) and 246.0 nm (5p $^2P_{1/2} \to$ 8s $^2S_{1/2}$) transitions in are presented Table 6.7, in addition to the 6s $^2S_{1/2}$ and 7s $^2S_{1/2}$ states, for comparison to literature transitions. The FS and MS factors for a transition are given by the difference between the FS or MS factors for the individual atomic states involved. The values from the FF and EVE approach calculations are also displayed in Table 6.7.

The normal mass shift can be approximated [54] by the non-relativistic expression $K_{NMS} = v_i \frac{m_e M_A}{m_e + M_A}$, where v_i is the frequency of the transition. M_A and m_e are the mass of the isotope and electron respectively, in units of u. These K_{NMS} values are shown in Table 6.7 as 'Energy' for comparison to the calculations, where the calculations determine the total isotope shift energy and then the K_{NMS} and K_{SMS} terms.

In order to determine the charge radii from the measured isotope shifts, one method typically employed is to use the King plot procedure as described in Sect. 2.2.3.3. In this work however, the King plot analysis was used to extract ratios of FS and MS factor for comparison to the calculations.

The King plots are shown in Fig. 6.18a for the $5p\,^2P_{3/2} \rightarrow 9s\,^2S_{1/2}$ (246.8 nm) and $5p\,^2P_{1/2} \rightarrow 8s\,^2S_{1/2}$ (246.0 nm) transitions and in Fig. 6.18b for the $5p\,^2P_{1/2} \rightarrow 6s\,^2S_{1/2}$ (410 nm) and $5p\,^2P_{3/2} \rightarrow 6s\,^2S_{1/2}$ (451 nm) transitions. The AR procedure calculation values are indicated by the blue lines in Fig. 6.18 and are within 1σ agreement of the experimental values. The FF values also agree within 1σ, while the EVE derived values result in an intercept on the King plot far outside of agreement, due to the significantly different MS factors.

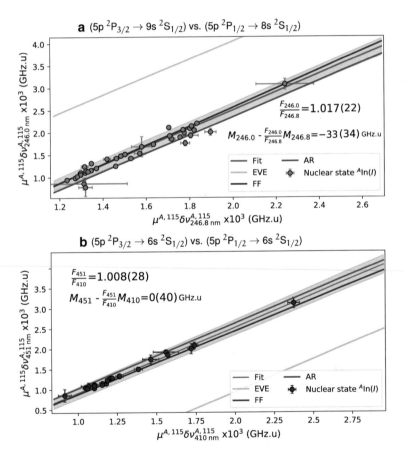

Fig. 6.18 King plots of the **a** 246.0 nm ($5p\,^2P_{1/2} \rightarrow 8s\,^2S_{1/2}$) and 246.8 nm ($5p\,^2P_{3/2} \rightarrow 9s\,^2S_{1/2}$) and **b** the 410.2 nm ($5p\,^2P_{1/2} \rightarrow 6s\,^2S_{1/2}$) and 451.1 nm ($5p\,^2P_{3/2} \rightarrow 6s\,^2S_{1/2}$) transitions. The shaded red area indicates the uncertainty of the linear fit

King plots between the combinations of transitions of this work (5p $^2P_{3/2} \rightarrow$ 9s $^2S_{1/2}$ (246.8 nm) and 5p $^2P_{1/2} \rightarrow$ 8s $^2S_{1/2}$ (246.0 nm)) and the transitions available from literature [36] (5p $^2P_{1/2} \rightarrow$ 6s $^2S_{1/2}$ (410 nm) and 5p $^2P_{3/2} \rightarrow$ 6s $^2S_{1/2}$ (451 nm)) were not used to extract the FS or MS parameters, as systematic errors (e.g.. voltage calibration) would not cancel when comparison is made between two experiments and would therefore introduce an error in the extracted King plot parameters. This is typically used to its advantage as the values from the King plot fix the $\delta \langle r^2 \rangle^{A,A'}$ extracted with a newly measured transition to agree with the $\delta \langle r^2 \rangle^{A,A'}$ values measured in another transition or with absolute $\langle r^2 \rangle^{A,A'}$ values from muonic data, for which the atomic parameters are already reliably known, despite any systematic error.

Orthogonal distance regression (ODR) [58] was used for the linear fit of the King plots. In the case of King plots ordinary least squares or distance regression fitting would no longer be valid as both the dependent and independent variables have measurement errors.

The measured isomer shift values, δv_i^m, to isomeric states, m, provide a test of the FS factors independent of the MS factor, as $\frac{A-A'}{AA'} \rightarrow 0$ for the isomeric states. Of the long-lived isomeric states in indium studied in this work, all have excitation energies below 1800 keV (most below 500 keV), which corresponds to an uncertainty of <0.05 MHz introduced by the assumption of $M_i \frac{A'-A}{A'A} = 0$. This is significantly smaller than the experimental uncertainties and is therefore neglected.

Therefore taking the ratio of the isomer shifts between the same states for two transitions then gives the ratio of the FS factors

$$\frac{\delta v_i^m}{\delta v_j^m} \cong \frac{F_i}{F_j} . \tag{6.15}$$

The fit to extract this ratio is shown in Fig. 6.19 for the available combinations of isomer shift measurements, δv_i^m, with the 5p $^2P_{3/2} \rightarrow$ 9s $^2S_{1/2}$ (246.8 nm) and 5p

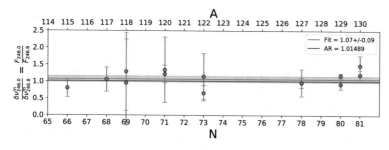

Fig. 6.19 The ratio of isomer shift values δv_i^m for the transitions 5p $^2P_{3/2} \rightarrow$ 9s $^2S_{1/2}$ (246.8 nm) and 5p $^2P_{1/2} \rightarrow$ 8s $^2S_{1/2}$ (246.0 nm) allowed determination of $F_{246.0}/F_{246.8} = 1.07(9)$ independent of the mass shift factor M_i

Table 6.8 The $\delta \langle r^2 \rangle^{113,115}$ values extracted from the $\delta v_i^{113,115}$ of the four optical transitions i. For comparison with $\delta \langle r^2 \rangle_{o\mu}^{113,115}$ extracted from muonic atom transitions [49, 57]

$$\frac{M_i^{Calc.} \mu^{113,115} - \delta v_i^{113,115}}{F_i^{Calc.}} = \delta \langle r^2 \rangle^{113,115} \ (\text{fm}^2)$$

Upper	$6s\ ^2S_{1/2}$	$6s\ ^2S_{1/2}$	$8s\ ^2S_{1/2}$	$9s\ ^2S_{1/2}$
Lower	$5p\ ^2P_{3/2}$	$5p\ ^2P_{1/2}$	$5p\ ^2P_{1/2}$	$5p\ ^2P_{3/2}$
AR	0.1512(3)	0.158(2)	0.162(3)	0.163(4)
FF	0.1465(3)	0.146(2)	0.149(3)	0.159(4)
EVE	0.279(4)	0.181(2)	0.180(4)	0.279(4)

$\delta \langle r^2 \rangle_{o\mu e}^{113,115}$ (fm^2) [57]		$\delta \langle r^2 \rangle_{\mu}^{113,115}$ (fm^2) [60]
0.157(11)		0.191(14)

$^2P_{1/2} \rightarrow 8s\ ^2S_{1/2}$ (246.0 nm) transitions used in this work, up to two δv_i^m values were measured at each isotope mass.

The literature measurements [36, 59] made with the 410.2 nm ($5p\ ^2P_{1/2} \rightarrow 6s\ ^2S_{1/2}$) and 451.1 nm ($5p\ ^2P_{3/2} \rightarrow 6s\ ^2S_{1/2}$) transitions did not have sufficient relative precision to report δv_i^m values. The extracted value of $F_{246.0}/F_{246.8} = 1.07(9)$ has 1σ agreement with both the AR calculated value of $F_{246.0}/F_{246.8} = 1.01489$ as well as the King plot value of $F_{246.0}/F_{246.8} = 1.017(22)$. The averaging would have to be performed with many more isomeric states, with improved precision on the isomer shift values, or between isomers with larger isomer shift values to determine the ratio $F_{246.0}/F_{246.8}$ more accurately.

Comparison to can be made to independent mean-square charge radii measurements of the stable ^{113,115}In isotopes, using Eq. 6.14, and represents a further test of the accuracy of the calculated atomic factors.

Two measurements of the charge radii difference, $\delta \langle r^2 \rangle_{\mu}^{113,115}$, have been reported in literature. In the most recent, using a combined analysis of muonic-atom and electron scattering data, [57] determined a value of $\delta \langle r^2 \rangle_{\mu}^{113,115} = 0.157(11)$ fm^2, while previously [60] reported a value of $\delta \langle r^2 \rangle_{\mu}^{113,115} = 0.191(14)$ fm^2 from muonic-atom measurements alone. The $\delta \langle r^2 \rangle_i^{113,115}$ values evaluated using the atomic factors from the AR, FF and EVE approaches are compared in Table 6.8. The AR approach produces values that have the best agreement with the literature values, within 1σ of the values reported by [57] and within 2σ of the work by [60]. Additionally the values using the AR procedure are consistent between all four transitions. The values extracted from the FF approach have a similar level of consistency between the transitions, but the values are underestimated compared to the literature values. While the values extracted using the EVE approach show the lowest level of consistency between the transitions, with two transitions 6σ overestimated compared to the closest literature value.

The SMS contribution of a state to the SMS of a transition decreases with increasing principle quantum of the state [61], therefore for transitions to high-lying Rydberg

Table 6.9 Comparison of experimentally evaluated SMS factors for individual atomic states s, $K_{SMS}^{Exp.+FS}$, extracted from experiment using Eq. 6.14 with calculated F_s factors, versus the calculated SMS factors, $K_{SMS}^{Calc.}$

	Method	AR		FF		EVE	
s	$\delta v_s^{113,115}$ (MHz)	$K_{SMS,s}^{Exp.+FS}$ (MHz)	$K_{SMS,s}^{Calc.}$ (GHz.amu)	$K_{SMS,s}^{Exp.+FS}$ (GHz.amu)	$K_{SMS,s}^{Calc.}$ (GHz.amu)	$K_{SMS,s}^{Exp.+FS}$ (GHz.amu)	$K_{SMS,s}^{Calc.}$ (GHz.amu)
$^2P_{3/2}$	272(6) [48]	−507(111)	−488	−462(114)	−403	69(101)	−899
$^2P_{1/2}$	277(10) [62]	−527(112)	−614	−525(119)	−470	−842(102)	−1048
$6s\,^2S_{1/2}$	17(6) [59]	168(50)	117	191(52)	119	68(50)	136
$7s\,^2S_{1/2}$	12(6) [63]	55(42)	39	67(42)	38	16(42)	42
$8s\,^2S_{1/2}$	9(12)	24(47)	16	31(48)	17	8(47)	18
$9s\,^2S_{1/2}$	2(10)	−13(43)	9	−14(43)	9	−22(43)	10

states the contribution from the upper state of the transition can be neglected. In order to test the SMS calculated for the individual atomic states s, the level isotope shift, $\delta v_s^{113,115}$, from the two-photon $5s^2 5p\,^2P_{3/2} \rightarrow 5s^2 np\,^2P_{1/2,3/2}$ $(n = 27\text{--}35)$ transitions [48] was used in combination with the other optical transition $\delta v_i^{113,115}$ values of this work, where $np\,^2P_{1/2,3/2}$ $(n = 27\text{--}35)$ are high-lying Rydberg states.

This allowed the determination of the level isotope shift values $\delta v_s^{113,115}$ values shown in Table 6.9 and the state-specific SMS values by Eq. 6.14, which are also displayed in Table 6.7. The subscript s in $\delta v_s^{113,115}$ indicates that the term is the isotope shift for the state not the transition. The agreement of the experimental values and the calculated values for the SMS is good using the AR approach, within 1σ for all atomic states. Agreement is also very good using the FF approach, with only the $6s\,^2S_{1/2}$ outside of 1σ agreement, while the EVE approach presents large discrepancies for the $^2P_{1/2}$, $^2P_{3/2}$ and $6s\,^2S_{1/2}$ states. The FS ratio using the EVE approach was similar to the values using AR or FF approaches in comparison to the values extracted from a King plot. However, the absolute values of the state FS have a large effect in the extraction of the experimental $K_{SMS,s}^{Exp+FS}$, and the values extracted using the EVE approach even changes the sign of the SMS, in addition to not agreeing with the calculated SMS values, $K_{SMS,s}^{Calc.}$. Greater precision and accuracy in the level isotope shift values would be needed to make any further comparison of the accuracy of the calculations for the other states using the FF approach.

6.5.2 Extracted Nuclear Charge Radii

The nuclear charge radii of $^{113-131}$In extracted with these new atomic factors from the AR approach are plotted in Fig. 6.20a for the two transitions of this work ($5p\,^2P_{3/2} \rightarrow 9s\,^2S_{1/2}$ (246.8 nm) and $5p\,^2P_{1/2} \rightarrow 8s\,^2S_{1/2}$ (246.0 nm)) in addition to

Fig. 6.20 **a** $\delta\langle r^2\rangle^{115,A}$ values for the $^{113-131}$In isotopes extracted from the IS measurements of four optical transitions and using the calculated FS and MS factors. Reference [36]—The error in extracting the FS from $\delta\langle r^2\rangle^{113,115}_{o\mu e}$ with range of $K_{SMS,451} = \pm2K_{NMS,451}$ is indicated by the yellow area. **b** $\sqrt{\langle r^2\rangle^A}$ compared to tin ($Z = 50$) [64] and cadmium ($Z = 48$) [65] isotopes

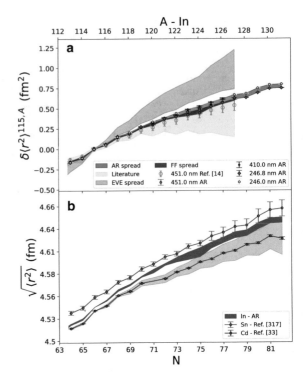

the two from literature (410.2 nm (5p $^2P_{1/2} \rightarrow$ 6s $^2S_{1/2}$) and 451.1 nm (5p $^2P_{3/2} \rightarrow$ 6s $^2S_{1/2}$)). The total spread between the evaluated $\delta\langle r^2\rangle^{115,A}$ values from the four atomic transitions are shown by the blue area in Fig. 6.20a. For comparison, the FF and EVE values are indicated by the purple and red areas respectively. The errors previously highlighted in the calculation of both the FS and MS factors using the EVE approach result in a large over estimation of the neutron-rich isotope radii and under estimation of the radii of the neutron-deficient isotopes.

As the FS and MS factors from the AR approach have passed experimental validation, one can have confidence in the accuracy of calculations and of the extracted nuclear charge radii values. This has allowed the nuclear charge radii of indium to be evaluated for the first time, independent of parameters from its neighbouring isotopes tin and cadmium. A plot of the absolute $\sqrt{\langle r^2\rangle^A}$ values is shown in Fig. 6.7b. The AR and FF values lie between the neighbouring isotopes of tin and cadmium, while the radii extracted using the EVE approach would be significantly larger than that for tin. Even considering the difference between all transitions used, the uncertainty on the AR or FF approach values is much smaller in the neutron-rich isotopes than the uncertainty introduced by previous FS and MS factors used in the cadmium [64] and tin [65] work. The $\delta\langle r^2\rangle^{115,A}$ values evaluated from these atomic factors are displayed in Table 6.10 (and plotted in Fig. 6.7a), which are used for the remainder of the discussion in this thesis.

Table 6.10 Compilation of the changes in mean-square charge radii extracted in this work using the 246.0 nm (5p $^2P_{1/2} \rightarrow$ 8s $^2S_{1/2}$) and 246.8 nm (5p $^2P_{3/2} \rightarrow$ 9s $^2S_{1/2}$) transitions. The systematic error from the AR approach calculations are stated inside the square brackets. The statistical errors are stated inside the round brackets

Mass	Spin	$\delta \langle r^2 \rangle^{115,A}$ 246.8 nm (fm^2)	$\delta \langle r^2 \rangle^{115,A}$ 246.0 nm (fm^2)
113	$\frac{9}{2}^+$	−0.163(4)[16]	−0.162(3)[17]
113	$\frac{1}{2}^-$	−0.139(5)[15]	−0.15(1)[19]
114	1^+	−0.115(5)[5]	
114	5^+	−0.103(7)[3]	−0.109(3)[6]
115	$^9/_2{}^+$	0	0
115	$\frac{1}{2}^-$	0.022(3)[1]	0.018(5)[1]
116	1^+	0.06(1)	
116	5^+	0.05(1)[1]	0.052(3)[1]
116	8^-	0.055(2)[6]	0.050(5)[6]
117	$\frac{9}{2}^+$	0.155(2)[18]	0.148(3)[17]
117	$\frac{1}{2}^-$	0.167(3)[17]	0.160(4)[16]
118	1^+	0.189(2)[18]	0.202(4)[16]
118	5^+	0.186(2)[18]	0.199(3)[17]
118	8^-	0.183(2)[18]	0.195(4)[16]
119	$\frac{9}{2}^+$	0.273(2)[28]	
119	$\frac{1}{2}^-$	0.282(3)[27]	
120	1^+	0.326(6)[36]	0.333(4)[36]
120	5^+	0.316(3)[37]	0.320(3)[37]
120	8^-	0.299(2)[38]	0.299(3)[37]
121	$\frac{9}{2}^+$	0.371(2)[48]	
121	$\frac{1}{2}^-$	0.376(2)[48]	
122	1^+	0.398(5)[45]	0.438(5)[55]
122	5^+	0.373(3)[47]	0.423(3)[57]
122	8^-	0.363(5)[45]	0.411(3)[57]
123	$\frac{9}{2}^+$	0.419(2)[58]	
123	$\frac{1}{2}^-$	0.415(2)[58]	
124	3^+	0.443(7)[63]	
124	8^-	0.444(2)[68]	
125	$\frac{9}{2}^+$	0.522(3)[67]	
125	$\frac{1}{2}^-$	0.511(4)[66]	
126	3^+	0.569(2)[78]	
126	8^-	0.564(3)[77]	
127	$\frac{9}{2}^+$	0.628(2)[88]	0.666(3)[88]
127	$\frac{1}{2}^-$	0.614(4)[86]	0.653(4)[86]
127	$\frac{21}{2}^-$	0.599(3)[87]	0.638(4)[86]
128	3^+	0.630(2)[88]	0.681(6)[94]

(continued)

Table 6.10 (continued)

Mass	Spin	$\delta\left\langle r^2\right\rangle^{115,A}$	$\delta\left\langle r^2\right\rangle^{115,A}$
		246.8 nm (fm²)	246.0 nm (fm²)
128	8^-	0.617(2)[88]	0.651(5)[95]
129	$\frac{9}{2}^+$	0.691(1)[99]	0.747(3)[97]
129	$\frac{1}{2}^-$	0.675(2)[98]	0.733(3)[97]
129	$\frac{23}{2}^-$	0.635(3)[97]	0.682(4)[96]
130	1^-	0.741(2)[98]	0.788(5)[95]
130	5^+	0.701(2)[98]	0.741(4)[96]
130	10^-	0.718(2)[98]	0.765(3)[97]
131	$\frac{9}{2}^+$	0.747(3)[97]	
131	$\frac{1}{2}^-$	0.749(2)[98]	0.795(6)[94]

6.6   Spin Assignments

6.6.1 $^{113-129}$In

The indium hyperfine spectra were initially fitted with the spin assignments from literature [66]. For the correct spin assignment the nuclear parameters in Eq. 2.26 should cancel for the upper and lower states when dividing by their respective hyperfine dipole constants, $A_{u,l}$, leaving only the atomic parameters, $J_{u,l}$ and $B_e(0)_{u,l}$, as

$$\frac{A_u}{A_l} = \frac{J_l\, B_e(0)_u}{J_u\, B_e(0)_l} \tag{6.16}$$

therefore consistency with this ratio can therefore be used to determine nuclear spin, I, given sufficient precision in the A_u/A_l ratio. This requires that one of the states in the atomic transition has $J > 1/2$ and it is not a $J = 0 \rightarrow 1$ transition. Therefore only the 5p $^2P_{3/2} \rightarrow$ 9s $^2S_{1/2}$ (246.8 nm) transition was sensitive to spin in this work. For all states with spins below $I = 4$ a clear change from $A_u/A_l = 0.543(3)$ was observed. This is shown in Fig. 6.21, showing a spin assignment of $I = 3$ to the ground state of ^{128}In. However for higher spins, although a minima in A_u/A_l was found for the literature spin assignments, and often also in the χ_R^2 of the fit, the error in A_u/A_l was too great to completely rule out neighbouring spins. See Fig. 6.24 for an example using the $I^\pi = 5^+$ and $I^\pi = 10^-$ structures in ^{130}In.

The A_u/A_l ratios for $^{113-131}$In are plotted in Fig. 6.22. The ratios are all in agreement of the ^{115}In reference isotope A_u/A_l ratio, with the following exceptions. For the $I^\pi = 1/2^-$ isomers the uncertainty on the A_u/A_l values is large, this is due to the small hyperfine splitting of the state and only two well resolved out of a total of three hyperfine transition peaks. However the unique number of peaks in the hyperfine structure rules out any spin assignment other than $I^\pi = 1/2$ for these isomers. The

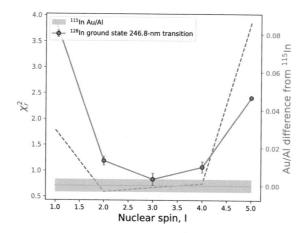

Fig. 6.21 The A_u/A_l ratio plot for the ground state in ^{128}In. For spins lower than $I = 4$ the A_u/A_l ratio allowed spin assignment. This is show in Fig. 6.22 for all isotopes

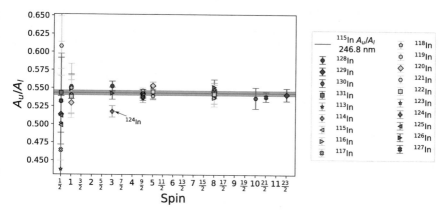

Fig. 6.22 The ratio between A_u and A_l should remain constant for correct spin assignments, given the hyperfine transition is sensitive to spin. The plot shows A_u/A_l for $^{113-131}$In for the $5p\,^2P_{3/2} \to 9s\,^2S_{1/2}$ (246.8 nm) transition

only other nuclear state not in agreement is the $I^\pi = 3^+$ of ^{124}In, but as mentioned in Sect. 6.3.2.1, this comes from a partial scan of the hyperfine structure. There was no evidence of an hyperfine anomaly outside the experimental uncertainty of this data, where the observed differential anomaly was of the order $^A\Delta^{A'} = -0.00238(13)\%$ for the $5p\,^2P_{3/2}$ state [67, 68] with the ^{113,115}In isotopes.

Also visible in Fig. 6.24 is the change in isomer shift with spin assignment. The difference in the field shift constants ratios $F_{246.8}/F_{246.0}$ for determining between two configurations of ^{130}In (Fig. 6.26) gave an incentive to test the spin assignments of $^{113-131}$In ratios for all isotopes using the approach described below.

As all masses measured had at least one isomer state, this allowed for a two- or three-point King plot type fit for each isotope mass to test for consistency with the $F_{246.8}/F_{246.0}$ ratio. The result of this analysis is shown in Fig. 6.26a. Where, for an example, the assignment with poorer agreement for ^{130}In is indicated as 'configuration B)' in the figure. Incidentally, ^{130}In is an ideal case for spin assignment using this method (the spin assignment of ^{130}In is discussed in the following sub section). The error on the $F_{246.8}/F_{246.0}$ value decreases with increasing isomer shift between the isomers, the precision on their centroids and the number of isomers. This may explain why this method has not to the authors knowledge been used before, as very few isotope chains have been measured with an abundance of isomers with a high precision. Those with smaller isomer shifts or only two nuclear states, such as ^{128}In have a large uncertainty on $F_{246.8}/F_{246.0}$ as the gradient is poorly defined, and so no firm conclusions can be made for these cases. However all spin assignments are consistent within 1, 2 σ using this method where possible for the $^{113-131}$In isotopes.

Using the additivity rule in Eq. 2.14 the calculated g-factors in Sect. 7.2 for the odd-odd masses are all consistent with measurement using literature spin assignments, except for the $I^\pi = 5^+$ isomer in ^{130}In, which is discussed in the following sub section. Additionally the quadrupole moments calculated with the additivity rule are also consistent, for all but the $I^\pi = 8^-$ states. The very constant magnetic moment of the $I^\pi = 9/2^+$ states, indicative of single-particle behaviour persists up until ^{129}In, supporting the assignment of this spin from the $g_{9/2}$ proton hole. The persistence of the accompanying $I^\pi = 1/2^-$ states also makes any other configuration for this single-hole state very unlikely.

6.6.2 ^{130}In Ambiguity

For the isotope ^{130}In, three isomeric states were known to exist from β^- and γ decay measurements [69]. Two were given unique spin assignments of $I^\pi = 1^-$ and $I^\pi = 5^+$ from β- and γ-ray selection rules and a tentative spin assignment of $I^\pi = 10^-$ for the third. The hyperfine structure spectrum of ^{130}In is shown in Fig. 6.23 for the 5p ^{2}P$_{1/2}$ → 8s ^{2}S$_{1/2}$ (246.0 nm) transition (used to show the issue clearly), the isotope was also measured with the 5p ^{2}P$_{3/2}$ → 9s ^{2}S$_{1/2}$ (246.8 nm) transition which gave the spin sensitivity.

The $I^\pi = 1^-$ state was easily identifiable as such, fitting with $I^\pi = 2$ resulted in unphysical isotope shifts ($\delta \langle r^2 \rangle = 0.5$ fm^2), changes in the A^u_{hf}/A^l_{hf} ratio and large χ^2_r values. However the two remaining hyperfine structures were not immediately identifiable as belonging to the $I^\pi = 5^+$ or $I^\pi = 10^-$ states, the magnetic moments extracted from each of the lines were consistent between the transitions.

Each of the hyperfine structures were fitted with a range of nuclear spins to cover both possibilities for each structure with the 5p ^{2}P$_{3/2}$ → 9s ^{2}S$_{1/2}$ (246.8 nm) transition. The trends are shown in Fig. 6.24 and suggest that the innermost structure in Fig. 6.23 belongs to the higher spin of the two, although both are within the

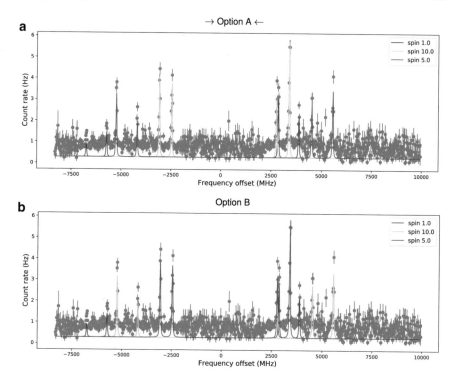

Fig. 6.23 Spectra of the two possible options for the assignment of hfs transitions to the higher spin isomers in ^{130}In, **a** the most intense isomer corresponds to $I^\pi = 10^-$ or **b** the most intense isomer corresponds to $I^\pi = 5^+$. Only spectra for the $5p\,^2P_{1/2} \to 8s\,^2S_{1/2}$ (246.0 nm) transition are shown for clarity

A^u_{hf}/A^l_{hf} ratio error and the hyperfine constants do not give a clear assignment to either.

Plotting the two options on a King plot as in Fig. 6.25 clearly reveals that only the assignment with $I^\pi = 10^-$ as the innermost hyperfine structure is compatible with the gradient of the King plot from the rest of the nuclear isotopes, option (A) in Fig. 6.23, in agreement with the A^u_{hf}/A^l_{hf} ratio plots. Therefore the innermost structure in Fig. 6.23 was assigned the high-spin configuration for the remainder of this work.

Although initially unexpected from a configuration point of view, the $I^\pi = 5^+$ assignment was assigned from β and γ spectroscopy, a spin of $I^\pi = 4^+$ was ruled out as an $E3$ transition could proceed to the $I^\pi = 1^-$ state [69].

The $F_{246.8}/F_{246.0}$ ratio of the two possible hyperfine structure configurations for ^{130}In are shown in Fig. 6.26a, indicating the agreement for 'configuration A)'. with the innermost structure being the higher-spin state. The agreement with the King plot $F_{246.8}/F_{246.0}$ ratio and rules out all configurations with $I \leq 7$ for the high-spin

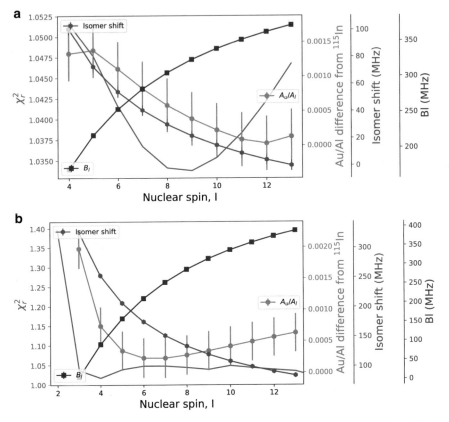

Fig. 6.24 Trends in the fitted hyperfine structure parameters with changing nuclear spin of the two structures shown in Fig. 6.23 for ^{130}In measure with the 5p ^{2}P$_{3/2}$ → 9s ^{2}S$_{1/2}$ (246.8 nm) transition. The parameters are shown for fitting to the **a** innermost and most intense structure and **b** the outermost and least intense structure

structure, with the best agreement for the assignment of $I = 10$ or $I = 9$, although $I = 8, 11, 12$ agreed within the uncertainty of the approach.

The spin assignments confirmed or found to be consistent with literature [66] using the above methods are summarised in Table 6.11.

6.7 Analysis of ^{131}In

The structure of the $I^\pi = 1/2^-$ isomeric state in ^{131}In is collapsed and systematically smaller than the previous neutron-rich isotopes $^{115-129}$In, which corresponds to a reduction in its magnetic moment at $N = 82$. This can be seen by comparing the spectra in Fig. 6.27a, b. The centroids of the $I^\pi = 9/2^+$ structure were used to combine

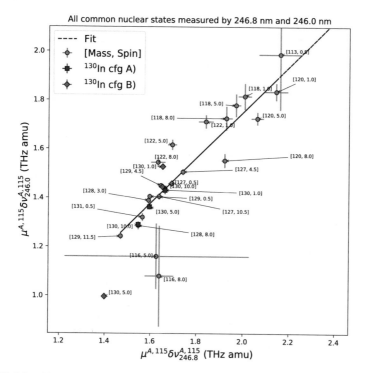

Fig. 6.25 Magnified section of King plot of all nuclear states measured with both 5p ^{2}P$_{3/2}$ → 9s ^{2}S$_{1/2}$ (246.8 nm) and 5p ^{2}P$_{1/2}$ → 8s ^{2}S$_{1/2}$ (246.0 nm) transitions. The two possibilities of spin assignments of the states in ^{130}In are shown in purple and red

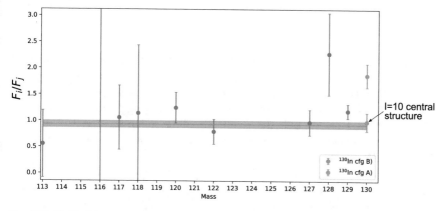

Fig. 6.26 Field shift constant ratios $F_{246.8}/F_{246.0}$ extracted from King plots of individual masses of the nuclear states measured with the 5p ^{2}P$_{3/2}$ → 9s ^{2}S$_{1/2}$ (246.8 nm) and 5p ^{2}P$_{1/2}$ → 8s ^{2}S$_{1/2}$ (246.0 nm) transitions. cfg (**a**) and cfg (**b**) refer to the hyperfine structure assignments shown in Fig. 6.23. The orange band indicates the $F_{246.8}/F_{246.0}$ extracted from the full King plot in Fig. 6.18

Table 6.11 Summary of the confidence in spin assignments of the states from this analysis compared to tentative assignments from literature [66]. 'g-factors' indicates consistency with the additivity rules discussed in Sect. 7.2. Inconsistency with the additivity rules may simply indicate the state is a mixed configuration different from the pure proton and neutron state coupling

Isotope mass	Proposed Spin	Conclusion
113–131 odd mass	$1/2^-$	Confirmed, hfs
113–131 odd mass	$9/2^+$	Consistent A_u/A_l, $F_{246.8}/F_{246.0}$
114–122 even mass	1^+	Confirmed, A_u/A_l
114–122 even mass	5^+	Consistent A_u/A_l, $F_{246.8}/F_{246.0}$, g-factors
126–128 even mass	3^+	Confirmed A_u/A_l, $F_{246.8}/F_{246.0}$, g-factors
114–128 even mass	8^+	Consistent A_u/A_l, $F_{246.8}/F_{246.0}$
127	$21/2^-$	Consistent A_u/A_l, $F_{246.8}/F_{246.0}$
129	$23/2^-$	Consistent A_u/A_l, $F_{246.8}/F_{246.0}$
130	1^-	Confirmed, A_u/A_l, g-factors
130	10^-	Consistent A_u/A_l, $F_{246.8}/F_{246.0}$. $12 \geq I \geq 7$, g-factors
130	5^+	Consistent A_u/A_l, $F_{246.8}/F_{246.0}$ $7 \geq I \geq 3$
		γ-, β-decay spectroscopy [69] $\neq 3^+, 4^+, 6^+$

three separate full scans of ^{131}In, in an attempt to make the central structure more immune to variations in wavelength error and better define the collapsed $I^\pi = 1/2^-$ peak structure. These three scans shared the same 5 proton pulses per laser frequency step, and therefore changes in background due to changing proton beam intensity should be small. However only the asymmetry of the collapsed $I^\pi = 1/2^-$ structure can be seen, which would give a large error on the μ value of this state with the 5p $^2P_{3/2} \rightarrow$ 9s $^2S_{1/2}$ (246.8 nm) transition alone.

The predicted centre of gravity of the ^{131}In isotope was scanned with the 5p $^2P_{1/2} \rightarrow$ 8s $^2S_{1/2}$ (246.0 nm) transition to give the spectrum shown in Fig. 6.28 of the $I^\pi = 1/2^-$ state. Fortunately the large A_{hf} of the 5p $^2P_{1/2} \rightarrow$ 8s $^2S_{1/2}$ (246.0 nm) transition allowed two of the three peaks for the $1/2^-$ to be resolved, providing a more precise determination of its magnetic dipole moment. The relative intensities of the smaller resolved peak and the collapsed doublet were consistent with the $I^\pi = 1/2^-$ assignment.

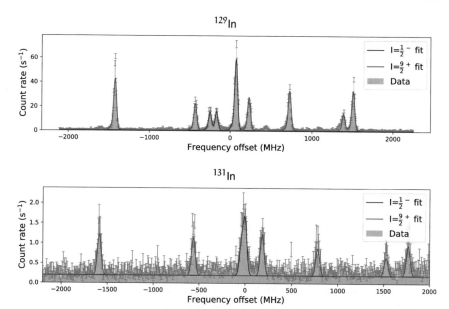

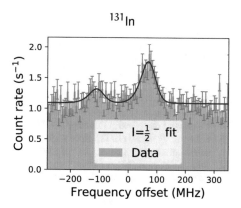

Fig. 6.27 **a** Spectrum of ^{129}In measured with the 246.8 nm (5p ^{2}P$_{3/2}$ → 9s ^{2}S$_{1/2}$) transition **b** Spectrum of ^{131}In measured with the 246.8 nm (5p ^{2}P$_{3/2}$ → 9s ^{2}S$_{1/2}$) transition

Fig. 6.28 Spectrum of ^{131}In measured with the 5p ^{2}P$_{1/2}$ → 8s ^{2}S$_{1/2}$ (246.0 nm) transition. Only the $I^\pi = {}^1/{}_2{}^-$ state was scanned

References

1. McKinney W (2010) In: Proceedings of the 9th python in science conference. pp. 51–56
2. HighFinesse (2019) Wavelength meter specifications. Technical report, HighFinesse.https://www.toptica.com/fileadmin/Editors_English/11_brochures_datasheets/03_HighFinesse/HighFinesse_Wavemeter.pdf
3. Morris MB et al (1984) Appl Op 23(21):3862. https://www.osapublishing.org/abstract.cfm?URI=ao-23-21-3862. ISSN 0003-6935
4. Kajava   TT   et   al   (1993)   J   Op   Soc   Am   B   10(11):1980. urlhttps://www.osapublishing.org/abstract.cfm?URI=josab-10-11-1980. ISSN 0740-3224

5. Steck DA (2001) Rubidium 87 D line data. Technical report, Los Alamos National Laboratory, Los Alamos, NM 87545. http://steck.us/alkalidata%5Cn, http://doi.wiley.com/10.1029/JB075i002p00463

6. Krieger A et al (2011) Nucl Instrum Methods Phys Res Section A: Accel Spectrom Detect Assoc Equip 632(1):23. https://www.sciencedirect.com/science/article/pii/S0168900210029256?via%3Dihub. ISSN 0168-9002

7. Pearson K (1895) Proc R Soc Lond 58(240–242)

8. Wang M et al (2017) Chin Phys C 41(3):30003. http://stacks.iop.org/1674-1137/41/i=3/a=030003?key=crossref.cbf64c2ee5251915f252c1f31e5f8e05. ISSN 1674-1137

9. Timar J et al (2014) Nucl Data Sheets 121:143. https://www.sciencedirect.com/science/article/pii/S0090375214006565. ISSN 0090-3752

10. Hashizume A (2011) Nucl Data Sheets 112(7):1647. https://www.sciencedirect.com/science/article/pii/S009037521100055X. ISSN 0090-3752

11. Newville M et al (2014) LMFIT: non-linear least-square minimization and curve-fitting for Python. Technical report, zenodo. https://zenodo.org/record/11813#.Wzecf6czbcs

12. Peterson EJ et al (2001) SciPy: open source scientific tools for Python. http://www.scipy.org/

13. Gins W et al (2018) Comput Phys Commun 222:286. https://doi.org/10.1016/j.cpc.2017.09.012. ISSN 00104655

14. Sonnenschein V (2015) Ph.D. thesis, Department of Physics, University of Jyvaskyla

15. Olivero J et al (1977) J Quant Spectrosc Radiat Transf 17(2):23. https://www.sciencedirect.com/science/article/pii/0022407377901613. ISSN 0022-4073

16. Halfmann T et al (2001) Opt Commun 199(November):117

17. Hong-fei JI et al, 2015009:1

18. Picqué J-L et al (1972) Opt Commun 5(5):402. https://www.sciencedirect.com/science/article/pii/0030401872900430. ISSN 0030-4018

19. Bendali N et al (1986) J Phys B: At Mol Phys 19(2):233. http://stacks.iop.org/0022-3700/19/i=2/a=012?key=crossref.0316417702b9b9d983b24575ea475f88. ISSN 0022-3700

20. James F (2006) Statistical methods in experimental physics. World Scientific

21. Ruiz RFG et al (2017) INTC-P5-04 proposal

22. Catherall R et al (2017) J Phys G: Nucl Particle Phys 44(9):094002.http://stacks.iop.org/0954-3899/44/i=9/a=094002?key=crossref.cb9ee490ac00453f72e007300938b2c6. ISSN 0954-3899

23. Thierry Stora KJ, Turrion M, Herman-Izycka U (2016) ISOLDE yield database. http://test-isolde-yields.web.cern.ch/test-isolde-yields/query_tgt.htm

24. Dillmann I et al (2002) Eur Phys J A 13(3):281. ISSN 14346001

25. Scheerer F et al (1992) Spectroch Acta Part B: At Spectrosc 47(6):793. https://www.sciencedirect.com/science/article/pii/058485479280074Q. ISSN 0584-8547

26. Jones MPA et al (2017) J Phys B: At Mol Opt Phys 50(6):060202. http://stacks.iop.org/0953-4075/50/i=6/a=060202?key=crossref.94a9f189f5b7eddcff03806f3e428c19

27. Sassmannshausen H et al (2016) Phys Rev Lett 117(8):083401.https://link.aps.org/doi/10.1103/PhysRevLett.117.083401. ISSN 0031-9007

28. Ovsiannikov VD et al (2011) Opt Spectrosc 111(1):25. http://link.springer.com/10.1134/S0030400X11070162. ISSN 0030-400X

29. Schulzt C et al (1991) J Phys B: At Mol Opt Phys J Phys B At Mol Opt Phys 24:48314844. http://iopscience.iop.org/article/10.1088/0953-4075/24/22/020/pdf

30. Aseyev SA et al (1993) Appl Phys B 56(1):391. http://aip.scitation.org/doi/abs/10.1063/1.47570 https://link.springer.com/content/pdf/10.1007%2FBF00324538.pdf. ISSN 0094243X

31. Koszorus A (2019) Nucl Instrum Methods Phys Res Sect B: Beam Interact Mater Atoms. Conference(EMIS2018)

32. Vernon A (2019) Nucl Instrum Methods Phys Res Sect B: Beam Interact Mater Atoms. EMIS2018:submitted

33. Nadeem A et al (2000) J Phys B: At Mol Op Phys 33(18):3729. http://stacks.iop.org/0953-4075/33/i=18/a=321?key=crossref.60386eeee9e4529543eb26ce458ce1e7. ISSN 0953-4075

34. Lynch KM et al (2014) Phys Rev X 4(1):1. http://link.aps.org/doi/10.1103/PhysRevX.4. 011055. ISSN 21603308

35. Fink DA et al (2015) Phys Rev X 5(1):011018. https://link.aps.org/doi/10.1103/PhysRevX.5. 011018. ISSN 2160-3308

36. Eberz J et al (1987) Nucl Phys A 464(1):9. http://linkinghub.elsevier.com/retrieve/pii/ 0375947487904192. ISSN 03759474

37. Eck TG et al (1957) Phys Rev 106(5):958. https://link.aps.org/doi/10.1103/PhysRev.106.958. ISSN 0031-899X

38. Flynn CP et al (1960) Proc Phys Soc 76(2):301. http://stacks.iop.org/0370-1328/76/i=2/a=415? key=crossref.f3515cb119eec3150e31da423e1fdb38. ISSN 0370-1328

39. Cameron JA et al (1962) Canad J Phys 40(8):931. http://www.nrcresearchpress.com/doi/10. 1139/p62-099. ISSN 0008-4204

40. Ionescu-Bujor M et al (1993) Hyperfine Interact 77(1):111. http://link.springer.com/10.1007/ BF02320305. ISSN 0304-3834

41. Childs WJ et al (1960) Phys Rev 118(6):1578. https://link.aps.org/doi/10.1103/PhysRev.118. 1578. ISSN 0031-899X

42. Nuytten C et al (1982) Phys Rev C 26(4):1701. https://link.aps.org/doi/10.1103/PhysRevC.26. 1701. ISSN 0556-2813

43. Lades H (1972) Zeitschrift für Physik A Hadrons Nucl 252(3):242. http://link.springer.com/ 10.1007/BF01395370. ISSN 0939-7922

44. Grupp H et al (1982) Nucl Phys A 386(1):56. https://www.sciencedirect.com/ science/article/abs/pii/0375947482904018, http://www.sciencedirect.com/science/article/pii/ 0375947482904018?via%3Dihub. ISSN 0375-9474

45. Antognini A (2015) URLhttp://arxiv.org/abs/1512.01765

46. Fricke G et al (2004) Introduction. Springer, Berlin/Heidelberg. http://materials.springer.com/ lb/docs/sm_lbs_978-3-540-45555-4_1

47. Fricke B et al (1977) At Data Nucl Data Tables 19(1):83. https://www.sciencedirect.com/ science/article/pii/0092640X77900109#!. ISSN 0092-640X

48. Menges R et al (1985) Z. Phys. A-At. Nucl. 320:575. https://link.springer.com/content/pdf/10. 1007%2FBF01411855.pdf

49. Jansen J (1989) Diploma thesis, University of Mainz

50. Fricke KHG (2004) Nuclear charge radii. Landolt-Börnstein - group I elementary particles, nuclei and atoms, vol. 20. Springer, Berlin/Heidelberg. http://materials.springer.com/bp/docs/ 978-3-540-45555-4

51. Sahoo BK et al (2020) New J Phys 22(1):012001. https://iopscience.iop.org/article/10.1088/ 1367-2630/ab66dd. ISSN 1367-2630

52. Monkhorst HJ (1977) Int J Q Chem 12(S11):421. http://doi.wiley.com/10.1002/qua. 560120850. ISSN 00207608

53. Bishop RF (1991) Theor. Chim. Acta 80(2–3):95. http://link.springer.com/10.1007/ BF01119617. ISSN 0040-5744

54. Bethe HA et al (1977) Quantum mechanics of one- and two-electron atoms. Springer, US. https://books.google.ch/books?id=RUfaBwAAQBAJ&dq=H.+A.+Bethe+and+E. +Salpeter+Quantum+Mechanics+of+One-+and+Two-Electron+atoms+Plenum/ Rosetta+New+York+1977&lr=&source=gbs_navlinks_s

55. Sahoo BK (2018) Private communication. Relativistic coupled cluster calculations for atomic states in indium

56. Kramida NATA, Ralchenko Y, Reader J (2014) NIST At Spect Database. http://www.nist.gov/ pml/data/asd.cfm

57. Fricke G et al (2004) 49-In Indium. Springer, Berlin/Heidelberg http://materials.springer.com/ lb/docs/sm_lbs_978-3-540-45555-4_51

58. Boggs PT et al (1990) Contemp. Math. 112:186 https://pdfs.semanticscholar.org/5749/ 2e2500b031fb19c2986544cbf6dcd4dabe74.pdf

59. Zaal GJ et al (1978) J Phys B: At Mol Phys 11(16):2821. http://stacks.iop.org/0022-3700/11/ i=16/a=009?key=crossref.c5c1bc343ddb14571dbb4db26f5a2de3. ISSN 0022-3700

60. Engfer R et al (1974) At Data Nucl Data Tables 14(5-6):509.https://www.sciencedirect.com/science/article/pii/S0092640X74800033. ISSN 0092-640X
61. Aldridge L et al (2011) Phys Rev A 84(3):034501. https://link.aps.org/doi/10.1103/PhysRevA.84.034501. ISSN 1050-2947
62. Garcia Ruiz RF et al. (2018) In: CERN-INTC-2018- P546. https://cds.cern.ch/record/2299760/files/INTC-P-546.pdf
63. Eliel ER et al (1981) Op Commun 36(5):366. https://www.sciencedirect.com/science/article/pii/0030401881902431. ISSN 0030-4018
64. Le Blanc FL et al (2005) Phys Rev C - Nucl Phys 72(3):1. ISSN 05562813
65. Hammen M et al (2018) Phys Rev Lett 121(10):102501. https://link.aps.org/doi/10.1103/PhysRevLett.121.102501. ISSN 0031-9007
66. Stone N (2005) At Data Nucl Data Tables 90(1):75. http://linkinghub.elsevier.com/retrieve/pii/S0092640X05000239. ISSN 0092640X
67. Eck TG et al (1957) Phys Rev 106(5):954. https://link.aps.org/doi/10.1103/PhysRev.106.954. ISSN 0031-899X
68. Persson JR (2013) At Data Nucl Data Tables 99(1):62. https://www.sciencedirect.com/science/article/pii/S0092640X1200085X. ISSN 0092640X
69. Fogelberg B et al (1985) Phys Rev C 31(3):1026. https://link.aps.org/doi/10.1103/PhysRevC.31.1026. ISSN 0556-2813

Chapter 7
Evolution of Neutron-Rich Indium Proton-Hole States

7.1 The Odd Indium Isotopes

7.1.1 Nuclear Magnetic Dipole Moments

The additivity rule, Eq. 2.14, can also be used to calculate the magnetic moment expected for a single nucleon with total angular momentum I in orbital L, with spin $S = 1/2$ as

$$\mu(I)_{s.p.} = I \left[\frac{1}{2}(g_L + g_S) + \frac{1}{2}(g_L - g_S)\frac{L(L+1) - \frac{3}{4}}{I(I+1)} \right], \qquad I = L \pm \frac{1}{2}$$

(7.1)

giving the single-particle values known as 'Schmidt' values [1]. See the Schmidt lines on Figs. 7.2 and 7.4 for comparison to the moments of the $I^\pi = 1/2^-$ and $I^\pi = 9/2^+$ states, with Schmidt values of $\mu(\pi 1g_{9/2}^{-1}) = +6.793\ \mu_N$ and $\mu(\pi 2p_{1/2}^{-1}) = -0.264\ \mu_N$ respectively.

Significant deviations from these single-particle Schmidt values are known to occur from residual interactions [2] and higher-order currents [3]. In this region of the nuclear chart deviations from the Schmidt values have been attributed to 'core polarisation' [4, 5] and meson-exchange currents (MECs) [6, 7] which modify the magnetic moments [8]. Chains of isotopes with closed shell Z configurations plus or minus a single proton offer a valuable probe of the coupling of the core and proton with varying neutron number and the dependence on the particle or hole nature of the proton. An explanation for the increase of the magnetic moment with neutron number of the antimony ($Z = 51$) $I^\pi = 7/2^+$ magnetic dipole moments [6], while the indium $I^\pi = 9/2^+$ ground state magnetic moments (Sect. 7.1.2) exhibit a fairly constant behaviour [9] is not yet clear. Furthermore, the mechanism for the decrease in the magnetic dipole moments of the indium isomer $I^\pi = 1/2^-$ states (Sect. 7.1.3) with increasing neutron number [9] is also not clear.

© The Editor(s) (if applicable) and The Author(s), under exclusive license
to Springer Nature Switzerland AG 2020
A. R. Vernon, *Collinear Resonance Ionization Spectroscopy
of Neutron-Rich Indium Isotopes*, Springer Theses,
https://doi.org/10.1007/978-3-030-54189-7_7

Residual interactions introduce configuration mixing [10], an effect attributed to excitations of the core (first-order configuration mixing is also known as 'core polarisation' [11]). The one-body μ operator in Eq. 2.13 can be modified into an *effective* operator by using effective g-factors, g_{eff}^{π}, which attempt to correct for neglected contributions [6, 12]. Many of these deviations can be due to neglected many-body contributions [13, 14] however, and an understanding of their origin may require ab-initio many-body calculations. Both coupled-cluster [15] and valence-space in-medium similarity-renormalization-group (VS-IMSRG) [13, 16] many-body methods are now able to reach the $Z = 50$ region of the nuclear chart.

7.1.2 The $I^{\pi} = {}^{9}/_{2}{}^{+}$ Single-Proton-Hole Ground States

The remarkably constant value of $\mu = 5.5 \pm 1$ μ_N of the $I^{\pi} = {}^{9}/_{2}{}^{+}$ ground states in indium, originally observed over 22 isotopes ${}^{103-127}$In [9, 17], has now been measured to persist up to ^{129}In. This constancy was thought to be due to the stability of the $1g_{9/2}^{-1}$ orbital configuration and core admixture, allowing a single-particle description to persist despite evolving neutron number. The constant value was described by a hole plus vibrating core model described by [18] using $g_s^{eff} = 0.7 g_s^{free}$, with the collective contribution to μ of less than 3% giving a final value of $\mu = 5.88$ μ_N [19]. A comparison with the new measurements for the μ values determined for the $I^{\pi} = {}^{9}/_{2}{}^{+}$ states is shown in Fig. 7.1.

The magnetic dipole moment for ^{129}In was found to be again consistent with this constant value only two neutrons away from the $N = 82$ shell closure However, for the ^{131}In isotope, the μ value abruptly increases towards the Schmidt value. This result contradicts the quenched g_s^{eff} single-particle picture with little influence from the neutron core.

Fig. 7.1 Nuclear magnetic dipole moments of the $I = {}^{9}/_{2}$ ground states extracted for the indium isotopes ${}^{113-131}$In. Hollow markers show the moment values extracted by Eberz87 [9]. Calculations by [20] are shown by the solid black line, with the individual contributions to the moment indicated

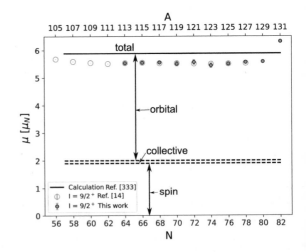

Fig. 7.2 Nuclear magnetic dipole moments of the $I = 9/2^+$ ground states extracted for the indium isotopes $^{113-131}$In. Hollow markers show the moment values extracted by [9]. Calculations by [21] with the phonon-coupling model are shown by the markers in red

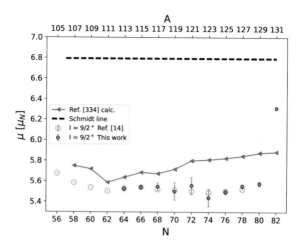

Collective modifications to the μ values were partially accounted for in the calculations by [21] including particle-phonon coupling [21]. The calculations by [21] used the Fayan's energy density functional [22] DF3-a, to create a self-consistent mean field model. The results are plotted alongside the experimental results of this work in Fig. 7.2. A gradual rise in the magnetic moment values towards the shell closure is predicted, which has poor agreement with the constancy of the values and sudden change at $N = 82$ observed by experiment.

All experimentally measured ground-state g-factors around neutron-rich shell closures to date are shown in Fig. 7.3a–c, for the $Z = 82$, $Z = 50$, and $Z = 20$ shell closures respectively. For single-proton hole isotopes with proton holes at $Z = 82$, $Z = 50$ and $Z = 20$ (other than potassium for which a spin change occurs at the shell closure (see Fig. 7.3c)), no experimental measurements of the magnetic moments have been made, and are challenging to access experimentally.

A linear increase with neutron number in the μ values of the antimony ($Z = 50 + 1$) ground states was observed, as seen in Fig. 7.3b. The μ values of ^{133}Sb [6] and ^{209}Bi [25] both were suggested to display evidence for meson-exchange current contributions [26], while the variation in the magnetic dipole moment with neutron number was attributed to collectivity, where an effective 'collective g-factor' was used to reproduce the slope. However there are no available measurements [6] of the Q_S values for antimony to assess the change in collectivity directly. The Tl ($Z = 81$) isotopes exhibit very constant magnetic dipole moment values of $\mu = 1.58 \pm 0.1 \, \mu_N$ up to the $N = 126$ shell closure, where a sudden increase is observed to $\mu = 1.876(5) \, \mu_N$ [23], seen in Fig. 7.3a. The origin of the ground states in thallium ($I = 1/2^+$) is from the proton hole $\pi 3s_{1/2}^{-1}$ orbital, which are highly susceptible to configuration mixing and for which meson-exchange currents contributions are suggested to be negligible [23, 27] (in contrast to the $I = 1/2^-$ states in indium discussed in Sect. 7.1.2). Additionally, it is not possible to isolate collective contributions [23] for the change in Tl, as the Q_S values cannot not be measured for spin $I = 1/2$ states, which is also an

Fig. 7.3 All experimentally measured ground-state g-factors of hole states around proton shell closures. **a** around $Z = 20 \pm 1$ (Sc and K [17]), **b** around $Z = 50 \pm 1$, (Sb [6]), **c** around $Z = 82 \pm 1$ (Bi [17] and Tl [17, 23, 24]). Note that the spin of the ground-state changes at $N = 28$, for potassium, and $N = 72$ for antimony. The uniqueness of the indium isotopic chain is discussed in the text

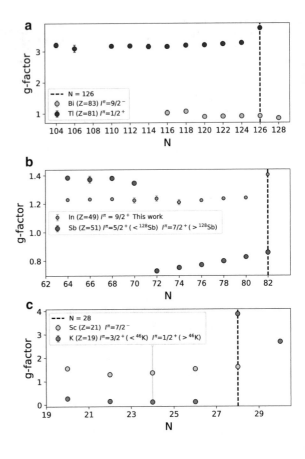

additional issue for the $I = 1/2^+$ spin state of ^{47}K at the $N = 26$ shell closure. These issues contrast with the $I^\pi = 9/2^+$ ground states in indium measured in this work.

The $1g_{9/2}^{-1}$ orbital $I^\pi = 9/2^+$ states of indium are therefore the only nuclear systems available to study the proton-hole-neutron interactions with the same orbital configuration up to a neutron-rich shell closure, below $Z = 82$. Additionally, the $I^\pi = 9/2^+$ states have observable Q_S values, which can be used as a measure of collectivity (see Sect. 7.1.4), and the accompanying $I^\pi = 1/2^-$ hole states (see Sect. 7.1.3) are uniquely positioned to isolate meson-exchange current or higher-order contributions to the μ values of the odd indium isotopes.

From comparison of this new experimental observation with the theoretical approaches to determine the μ values mentioned above in Figs. 7.2 and 7.1, it is clear that they are missing an underlying contribution which would predict the process occurring at the $N = 82$ closure for indium.

7.1.3 The $I^\pi = 1/2^-$ Single-Proton-Hole Isomer States

A clear deviation in the μ values from the Schmidt value for the $I^\pi = 1/2^-$ states can be seen in Fig. 7.4. The decrease in μ with neutron number and crossing of the Schmidt value at ^{117}In was originally observed by [9], while the increase after $N = 76$ is a new observation. Attempts to explain this behaviour in terms of configuration mixing with collective excitations would be inconsistent with the observed stability of the $I^\pi = 9/2^+$ moments [19]. The deviation has remained an unresolved problem in the interpretation of nuclear structure in this region of the nuclear chart.

Admixtures of other configurations can contribute to cause deviations from the Schmidt values [28]. In the configuration mixing picture the final nuclear wave function $|\psi_I\rangle$ can be written [29] as

$$|\psi_I\rangle \propto |\chi_I\rangle + \sum_P \alpha_P |\phi_{P,I}\rangle \, , \tag{7.2}$$

where $|\chi_I\rangle$ represents the extreme single-particle configuration, and $|\phi_{P,I}\rangle$ represents the wave functions of possible admixed configurations, P, which can combine to I, denoting the nuclear spin of the state of interest. Admixtures linear in α_P are significant, but this requires a maximum difference of one particle state and that the orbital state is the same, $l + 1/2$ to $l - 1/2$ excitations satisfy this condition [30]. However for $p_{1/2}$ nuclei first-order configuration mixing contributions go to zero for excitations of a proton or neutron from the core [29]. This can be seen in the deviations of magnetic moments, $\delta\mu$, derived for first-order configuration mixing by [10]. For an odd proton number in a $j = l - 1/2$ orbital, the deviation has the coefficient

$$\delta\mu \propto \frac{(l-1)l_1}{(2l+1)(2l_1+1)} \, , \tag{7.3}$$

Fig. 7.4 Nuclear magnetic dipole moments of the $I^\pi = 1/2^-$ isomer states of the indium isotopes $^{113-131}$In. Hollow markers show the moment values extracted by [9]

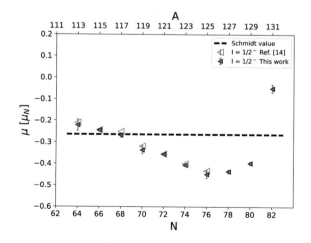

where l_1 is the valance nucleon orbital angular momentum i.e. $l_1 = l$. $\delta\mu$ becomes zero for the $2p_{1/2}$ proton orbital and therefore for the $I^\pi = 1/2^-$ states. The collective contribution to the change in the μ values of the $I^\pi = 1/2^-$ states are expected to be small, as they would also affect the $I^\pi = 9/2^+$ μ values which are observed to be constant up to $N = 80$ (see also the isomer shift discussion in Sect. 7.3.2). This suggests that MECs or higher-order configuration mixing could add an appreciable contribution to the deviation of the $I^\pi = 1/2^-$ states from the Schmidt line.

7.1.4   The $I^\pi = 9/2^+$ State Nuclear Electric Quadrupole Moments

The particle-core coupling model of [18] predicts a large collective contribution to the $I^\pi = 9/2^+$ state quadrupole moments, Q_S, of ~60%. Values of $Q^{eff} \approx 0.7\ b$ with a decrease towards $N = 78$ were calculated by [19], as shown in Fig. 7.5. This decrease is observed to continue, with a remaining deformation at the $N = 82$ shell closure, which may be due to a remaining core polarising effect of the proton hole. From these Q_S values alone it is not possible to discern any collective contributions which would cause the sudden increase in the μ value at the $N = 82$ shell closure.

The phonon-coupling calculations by [31] were also performed for the $I^\pi = 9/2^+$ quadrupole moments. The results are shown in red in Fig. 7.5. While the agreement is poor with experiment, the general trend towards the shell closure is reproduced. The dip in Q values around $N = 64$ measured by experiment could be indicative of a sub-shell closure, this is a potential reason for the large deviation of the Q experimental values towards the mid-shell compared to calculations by [31].

While neither theories mentioned above were able to reproduce the measured Q_S values within experimental error, the stability at the mid-shell was reproduced in the

Fig. 7.5 Nuclear electric quadrupole moments of the $I = 9/2$ ground states extracted for the indium isotopes $^{113-131}$In. Hollow markers show the moment values extracted by [9]. Calculations by [31] with the phonon-coupling model are shown by the hollow markers in red. Calculations by [20] shown in black. See Fig. 7.12 for the determined static deformation parameters, β_2^{static}

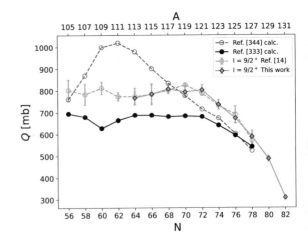

calculations by [19]; this may be due to the inclusion of all proton orbitals between $Z = 28$ and $Z = 50$.

Shell-model calculations and measurements of the B(E2) values in the tin isotopes similarly display a parabolic behaviour around $N = 66$ [32–36]. The increase in deformation with a maximum at the mid-shell is expected from the seniority scheme [37–39]. The saturation of deformation at the mid-shell of the even-mass tin isotopes has been attributed to a collectivity reducing effect of the $\nu 3s_{1/2}$ orbital [35]. This coincides with the ground state spin of the tin isotopes becoming $I^{\pi} = {}^{1}\!/_{2}{}^{+}$ for ${}^{113-119}$Sn and $I^{\pi} = 5^{+}$ $(\pi 1g_{9/2}^{-1} \otimes \nu 3s_{1/2})$ for ${}^{114-122}$In.

A comparison of the static deformation determined from these Q_{S} values to the total deformation parameters extracted from charge radii is made in Sect. 7.3.1 for these $I^{\pi} = {}^{9}\!/_{2}{}^{+}$ states. The accurate calculation of the evolution of the Q_{S} values towards $N = 82$, and of the measured value at ^{131}In may give a better understanding of the important contributions to collectivity in the $Z = 50$ region.

7.2 The Even Indium Isotopes

7.2.1 Nuclear Magnetic Dipole Moments

The magnetic dipole moment additivity rule shown in Eq. 2.14 was used to calculate the composite moments of the states in the odd indium isotopes. This also helped to give further evidence of the assigned spins to the states, which were so far found to be consistent with literature [17], and indicates the purity of the single-particle configurations. Values from tin [17, 40] were used for the relevant neutron I_{n} states and indium for the I_{p} proton states from the odd indium isotopes of this work. The assumed coupling between configurations which were used are shown in Table 7.1.

The resultant g-factors from these magnetic moment couplings are shown in Fig. 7.6. The additivity-rule values slightly overestimates the g-factors for the 1^{+}, 3^{+} and 5^{+} states, but given no residual nuclear effects are included this gives confidence in the coupling configurations used.

The $I^{\pi} = 8^{-}$ states show an increasing departure from their additivity rule values with neutron number, while the $I^{\pi} = 5^{+}$ isomer in ^{130}In has the biggest departure from the additivity rule values for all the states. Using the coupling $1g_{9/2}^{-1} \otimes \nu 1h_{11/2}$; 5^{-} instead gives much better agreement for the magnetic moment ($g = 0.52$) but this would predict a change in sign for the quadrupole moment ($Q_{S} = -172$ mb), which was not observed and can therefore be ruled out. The $I^{\pi} = 5^{+}$ isomer of the ^{130}In isotope appears from the $1g_{9/2}^{-1} \otimes \nu 2d_{3/2}$ configuration coupling, it is not immediately apparent why this results in a spin of 5^{+} rather than 3^{+} or 6^{+}. The existence of the 5^{+} isomer spin over 3^{+} or 6^{+} from the $1g_{9/2}^{-1} \otimes \nu 2d_{3/2}$ configuration coupling is not obvious. It has been suggested [41] that this might be due to a mixing contribution from the $1g_{9/2}^{-1} \otimes \nu 3s_{1/2}$ configuration making the $I^{\pi} = 5^{+}$ from the $1g_{9/2}^{-1} \otimes \nu 2d_{3/2}$ configuration preferable. In this case it is not surprising that

Table 7.1 Primary configuration couplings expected to produce the nuclear ground and isomeric spin states in $^{113-131}$In

Odd-Z Even-N			Odd-Z Odd-N		
Mass	Spin	Configuration	Mass	Spin	Configuration
113	$9/2^+$	$\pi 1g_{9/2}^{-1}$	114	1^+	$\pi 1g_{9/2}^{-1} \otimes \nu 1g_{7/2}$
113	$1/2^-$	$\pi 2p_{1/2}^{-1}$	114	5^+	$\pi 1g_{9/2}^{-1} \otimes \nu 3s_{1/2}$
115	$9/2^+$	$\pi 1g_{9/2}^{-1}$	116	1^+	$\pi 1g_{9/2}^{-1} \otimes \nu 1g_{7/2}$
115	$1/2^-$	$\pi 2p_{1/2}^{-1}$	116	5^+	$\pi 1g_{9/2}^{-1} \otimes \nu 3s_{1/2}$
117	$9/2^+$	$\pi 1g_{9/2}^{-1}$	116	8^-	$\pi 1g_{9/2}^{-1} \otimes \nu 1h_{11/2}$
117	$1/2^-$	$\pi 2p_{1/2}^{-1}$	118	1^+	$\pi 1g_{9/2}^{-1} \otimes \nu 1g_{7/2}$
119	$9/2^+$	$\pi 1g_{9/2}^{-1}$	118	5^+	$\pi 1g_{9/2}^{-1} \otimes \nu 3s_{1/2}$
119	$1/2^-$	$\pi 2p_{1/2}^{-1}$	118	8^-	$\pi 1g_{9/2}^{-1} \otimes \nu 1h_{11/2}$
121	$9/2^+$	$\pi 1g_{9/2}^{-1}$	120	1^+	$\pi 1g_{9/2}^{-1} \otimes \nu 1g_{7/2}$
121	$1/2^-$	$\pi 2p_{1/2}^{-1}$	120	5^+	$\pi 1g_{9/2}^{-1} \otimes \nu 3s_{1/2}$
123	$9/2^+$	$\pi 1g_{9/2}^{-1}$	120	8^-	$\pi 1g_{9/2}^{-1} \otimes \nu 1h_{11/2}$
123	$1/2^-$	$\pi 2p_{1/2}^{-1}$	122	1^+	$\pi 1g_{9/2}^{-1} \otimes \nu 1g_{7/2}$
125	$9/2^+$	$\pi 1g_{9/2}^{-1}$	122	5^+	$\pi 1g_{9/2}^{-1} \otimes \nu 3s_{1/2}$
125	$1/2^-$	$\pi 2p_{1/2}^{-1}$	122	8^-	$\pi 1g_{9/2}^{-1} \otimes \nu 1h_{11/2}$
127	$9/2^+$	$\pi 1g_{9/2}^{-1}$	124	3^+	$\pi 1g_{9/2}^{-1} \otimes \nu 2d_{3/2}$
127	$1/2^-$	$\pi 2p_{1/2}^{-1}$	124	8^-	$\pi 1g_{9/2}^{-1} \otimes \nu 1h_{11/2}$
127	$21/2^-$	$\pi 1g_{9/2}^{-1} \otimes \nu 1h_{11/2} \otimes \nu 2d_{3/2}^{-1}$	126	3^+	$\pi 1g_{9/2}^{-1} \otimes \nu 2d_{3/2}$
129	$9/2^+$	$\pi 1g_{9/2}^{-1}$	126	8^-	$\pi 1g_{9/2}^{-1} \otimes \nu 1h_{11/2}$
129	$1/2^-$	$\pi 2p_{1/2}^{-1}$	128	3^+	$\pi 1g_{9/2}^{-1} \otimes \nu 2d_{3/2}$
129	$23/2^-$	$\pi 1g_{9/2}^{-1} \otimes \nu 1h_{11/2} \otimes \nu 2d_{3/2}^{-1}$	128	8^-	$\pi 1g_{9/2}^{-1} \otimes \nu 1h_{11/2}$
131	$9/2^+$	$\pi 1g_{9/2}^{-1}$	130	1^-	$\pi 1g_{9/2}^{-1} \otimes \nu 1h_{11/2}$
131	$1/2^-$	$\pi 2p_{1/2}^{-1}$	130	10^-	$\pi 1g_{9/2}^{-1} \otimes \nu 1h_{11/2}$
			130	5^+	$\pi 1g_{9/2}^{-1} \otimes \nu 2d_{3/2}$

these simple additivity rules do not agree with the experimental moments for the $I^\pi = 5^+$ state. In combination with the relative pure 1^- and 10^- states from $\pi 1g_{9/2}^{-1} \otimes \nu 1h_{11/2}$ aligned and anti-aligned configurations, the simultaneous reproduction of the μ values for the three isomer states in ^{130}In by nuclear theory may give a further explanation into the preference for the $I^\pi = 5^+$ spin state.

The additivity rule predicts a constant μ value with increasing neutron number for the $I^\pi = 8^-$ states. The measured magnetic moments are shown in Fig. 7.7a. The

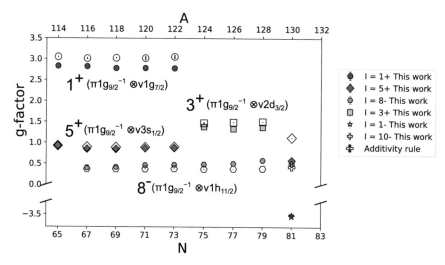

Fig. 7.6 The g-factors of the odd indium isotopes moments extracted using the A_{lower} of the measured $5p\,^2P_{3/2} \rightarrow 9s\,^2S_{1/2}$ (246.8-nm) transition, compared to those predicted by the two-particle additivity rule. The error bar on the additivity rule values is only from the single-particle values used

values increase abruptly after $N = 70$ and again after $N = 78$ with a linear increase with neutron number between these jumps. This second jump at ^{128}In was not previously observed. It has been suggested that a possible cause for these deviations could be due to neutron pairs filling orbitals out of order due to pairing correlations [19]. Measurements of the spin of the corresponding even-odd tin isotopes corroborate this idea [17], where the ground-state spin goes from $I^\pi = 1/2^+$ for $N = 63$–69 then becomes $I^\pi = 3/2^+$ at $N = 71$, then turns to $I^\pi = 11/2^-$ for $N = 73$–77 before finally returning to $I^\pi = 3/2^+$ for $N = 79$–81. In recent measurements of the spins and moments of $^{107-129}$Cd [43] the only deviation from an otherwise linear dependence of the observed μ values is over $N = 73$–81 with a downwards dip of $\mu \approx 0.2\,\mu_N$. This corresponds to a change in the ground-state spin from $I^\pi = 1/2^+$ for $N = 63$–71 to $I^\pi = 3/2^+$ for $N = 73$–81. The trend in the $I^\pi = 8^-$ states of indium, which results from the coupling of this $\nu 1h_{11/2}$ neutron orbital with the seemingly stable $(\pi g_{9/2})^{-1}$ states from the proton hole, is not a constant independent of neutron number as the additivity rule would predict.

In Fig. 7.7b, a comparison is made to the $I^\pi = 8^-$ values determined by the additivity rules, using the cadmium neutron-state μ values [43] rather than tin [44]. The experimentally determined μ values from fitting with spins of $I = 7$, $I = 9$ and $I = 10$ for the high-spin state in ^{128}In and the corresponding additivity rule values are indicated. The measured and additivity rule values for the $I^\pi = 10^-$ state in ^{130}In are also shown. The use of the cadmium values improves the agreement with the ^{116,118}In μ values, however the step changes are not reproduced with the additivity rule, and therefore spin assignments based on the values is not appropriate.

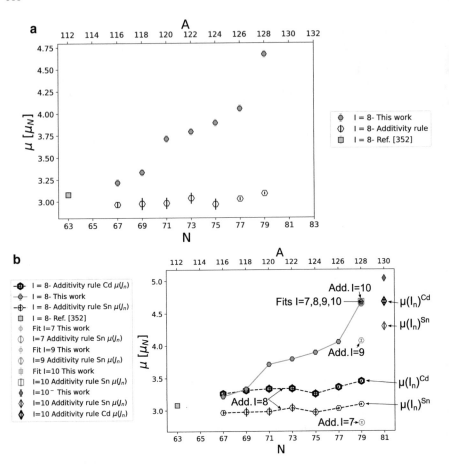

Fig. 7.7 **a** The magnetic dipole moments of the $I^{\pi} = 8^{-}$ states extracted using the A_{lower} of the measured 5p ^{2}P$_{3/2}$ → 9s ^{2}S$_{1/2}$ (246.8-nm) transition, compared to those predicted by the two-particle additivity rule. Literature value of ^{112}In [42]. **b** Comparison of the difference between additivity rule values using cadmium and tin reference neutron-state magnetic moment values, $\mu(J_{n})$. The additivity rule values for $I = 7, 9, 10$ are also indicated alongside the determined experimental μ values fitting with spins $I = 7, 9, 10$. The same comparison is shown for the $I^{\pi} = 10^{-}$ state in ^{130}In

Spin assignments of $I = 7$ to $I = 10$ were consistent with the analysis in this work for the tentatively assigned [45, 46] spin $I = 8$ states of ^{120,122,124,126}In. While the additivity rule could indicate that the spin changes to $I = 9$ for ^{120,122,124,126}In and then $I = 10$ for ^{128,130}In, the difference between additivity rule values using tin or cadmium moment values for the $I = 8$ and $I = 10$ assignments of $^{116-128}$In and ^{130}In respectively, indicate there are underlying structural changes which are not accounted for by the additivity rule.

In contrast with these $I^{\pi} = 8^{-}$ states in indium and the cadmium ground states, the magnetic moments of tin display very little variation with neutron number [40].

This is expected in the absence of core polarisation [47]. The two proton hole in cadmium has a significant effect on the core [43] nucleons, resulting in the linear dependence with neutron number, the change in ground-state spins, and the deviation from the linear dependence coinciding with this change in spin.

It appears that there may be a change in the core structure correlated with the change in spin to $I^\pi = 3/2^+$ at $N = 71$ in tin then occurs at $N = 71$–79 in indium and $N = 73$–81 in cadmium. Any model describing these states in the indium isotopes will need to include higher order effects, which are not accounted for by these simple additivity rules. This may indicate the effect of a single-proton hole is significant on the underlying neutron structure, and may principally be due to the effect of pair correlations which lead to changes in core polarisation. Recent Monte-Carlo configuration interaction calculations by [36] for tin also motivate this interpretation, they predict a phase change in neutron separation energy which finishes at $N = 70$; this prediction is due to pair correlations which take over at $N = 66$.

Further measurements which confirm the spins unambiguously or calculations which reproduce the step changes seen in the $I^\pi = 8^-$ states may shed light on their origin (in addition to the yet unaddressed deviation seen in the $h_{11/2}$ magnetic dipole moments of the cadmium isotopes [43]).

7.2.2 Nuclear Electric Quadrupole Moments

The extracted quadrupole moments from this work are plotted in Fig. 7.8 alongside literature values where available. Ab-initio electric field gradient calculations by [48] were applied to a quadrupole coupling constant measured by a solid-state experiment using the nuclear quadrupole alignment technique [49] to extract a quadrupole moment for the 5^+ state in ^{114}In of $Q(^{114}$In$) = -0.14(1)$ b. This was in clear disagreement with the value of $+0.7(1)$ b measured by [9] using collinear laser spectroscopy, the proposed reason was that the electric field gradient calculation in [9] used approximations including 'Sternheimer corrections' [50]. For indium the expected order of these corrections is only around 10% however [50]. The measurement of ^{114}In in this work combined with accurate calculation of the electric-field gradient for the free indium atom (Sect. 5.5) yields a value of $Q(^{114}$In$) = +0.716(22)$ b. Given the repeated and consistent measurement of this value, and the increased complexity of both experiment measurements and theoretical calculations of electric field gradients within a solid compared to an atom for collinear laser spectroscopy, it is likely that either the experimental value input [49] or the calculations performed by [48] contain an error which resulted in their value of $Q_S(^{114}$In$) = -0.14(1)$ b.

The quadrupole moments of the 5^+ states in ^{116}In and ^{118}In reported in this thesis are not in agreement with literature value of [9], however the work does not state a systematic error for their values and the ^{114}In and ^{104}In Q_S values for the 5^+ states also measured with the same transition have significantly larger error bars. The 5^+ quadrupole values determined in this work will be used for consistency.

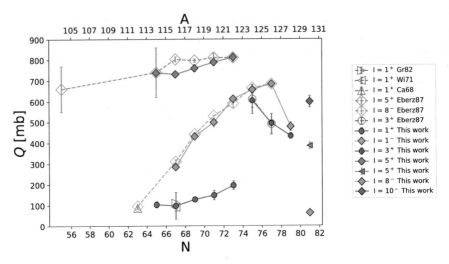

Fig. 7.8 The nuclear spectroscopic-quadrupole moments of the odd indium isotopes moments extracted using the B_{lower} of the measured $5p\,^2P_{3/2} \to 9s\,^2S_{1/2}$ (246.8-nm) transition. Literature values were taken from [9, 51–53]

As with the magnetic dipole moments, the additivity rules give fair agreement for the 1^+, 3^+ and 5^+ states but not for the 8^- or ^{130}In 5^+ state. It is suggestive of shared underlying mechanism for the deviations of the Q_S values of the 1^+ and 5^+ states given their agreement within 10 mb given a constant offset of 600 mb between them over $^{114-122}$In. The overestimation of the Q_S for the 10^- state is also large here. Collective effects are known to play a role in these high-spin states however, but are neglected by this simple rule [54, 55].

The deviation of the $I^\pi = 8^-$ states from that expected for the additivity rule was previously seen up to ^{126}In [9], and could also follow from the pair correlations which were discussed for the magnetic moments. At ^{128}In, the Q_S value is now observed to suddenly decrease while its μ value increases. This is consistent with the idea that the neutron core structure is changing over $N = 71–79$ in indium and once the orbital filling is over, the Q_S value drops as the effective charge from this core polarisation reduces (Fig. 7.9).

7.3   Nuclear Charge Radii of the Indium Isotopes

The atomic factors used in this work calculated by the AR approach (see Sect. 6.5.1) have allowed evaluation of the nuclear charge radii, $\sqrt{\langle r^2 \rangle^A}$, of indium for the first time, independent of its neighbouring isotones. The value of $\sqrt{\langle r^2 \rangle^A} = 4.615$ fm [56] was used for the ^{115}In reference isotope. These values are compared to its

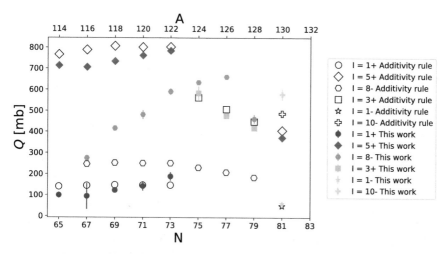

Fig. 7.9 The nuclear spectroscopic-quadrupole moments of the odd indium isotopes moments extracted using the B_{lower} of the measured 5p $^2P_{3/2} \rightarrow$ 9s $^2S_{1/2}$ (246.8-nm) transition, compared to those predicted by the two-particle additivity rule

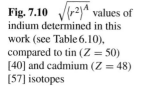

Fig. 7.10 $\sqrt{\langle r^2 \rangle^A}$ values of indium determined in this work (see Table 6.10), compared to tin ($Z = 50$) [40] and cadmium ($Z = 48$) [57] isotopes

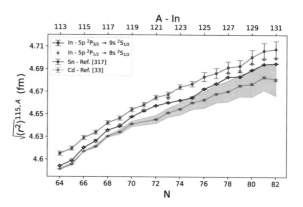

neighbouring isotopes tin ($Z = 50$) [40] and cadmium ($Z = 48$) [57] in Fig. 7.10. An interesting trend can be seen, the radii of the indium isotopes have a similar charge radius to tin on the neutron rich side, while they are similar to cadmium for the proton rich. Although the trend could be due to a deficiency in the new atomic factor calculations, or those used in the reported values of cadmium (for which the uncertainties are large towards neutron-rich) and tin, similar trends can be seen in other regions of the nuclear chart [58]. Another feature to note is the kink at $N = 79$ for both the indium and tin isotopes, but not the cadmium. This feature is clearer in the evaluated odd-even staggering of the charge radii, to be discussed below (Sect. 7.3.3).

7.3.1 Deformation

The determined $\delta \langle r^2 \rangle^A$ values of $^{113-131}$In are plotted against isodeformation lines from the droplet model [59, 60] in Fig. 7.11. This indicates that ^{131}In has a remaining deformation contribution to its $\langle r^2 \rangle^A$ value. Although this is a highly nuclear model dependent reference for a spherical nucleus, the droplet model with no deformation agrees with the experimental value of ^{132}Sn [40]. The deformation is observed to increase towards the mid-shell, reaching a deformation $\beta_2^{total} \sim 0.25$ below N = 72. This can be seen in comparison to the static deformation parameters, β_2^{static}, extracted from the Q_S values of the $I^\pi = 9/2^+$ states in Fig. 7.12. The dynamic deformation parameter, $\beta_2^{dynamic}$, is also indicated (from Eq. 2.22). Despite the model dependent offset in the β_2^{total} values, it is clear that the dynamic contribution to the deformation increases significantly at the mid-shell. The parabolic behaviour in the β_2^{total} values determined from the nuclear charge radii is predicted by the seniority scheme [5]. However it is not possible to pinpoint the origin of the large dynamic contribution from these observations alone. Variations in surface thickness with neutron number [61] or dynamic quadrupole and octupole contributions [62] could play a role in explaining the increased charge radii outside of the β_2^{static} contribution. The β_2^{total} values for the $I^\pi = 1/2^-$ states are also indicated in Fig. 7.12. The difference from in the β_2^{total} values from the $I^\pi = 9/2^+$ states is very small on this scale and indicates that the dynamic and static deformations may be shared between the states (see Sect. 7.3.2).

Fig. 7.11 Changes in mean-square nuclear charge radii of the $^{113-131}$In isotopes determined from measurements with the $5p\,^2P_{3/2} \rightarrow 9s\,^2S_{1/2}$ (246.8-nm) transition. Isodeformation lines determined from the droplet model are also shown [59, 60]

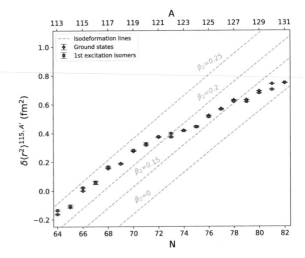

Fig. 7.12 Evolution of the deformation parameters, β_2, for the $I^\pi = 9/2^+$ states. The β_2^{static} values were determined from the Q_S values and Eqs. 2.18 (giving Q_0) and 2.21. The β_2^{total} values from the $\delta\langle r^2\rangle^{115,A'}$ values and Eq. 2.11

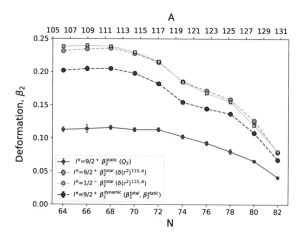

7.3.2   Isomer Shifts

The shift in charge radius of the isomer states in the odd $^{113-131}$In isotopes are shown relative to the $I^\pi = 9/2^+$ ground states in Fig. 7.13a. These isomer shifts were not previously measured with sufficient precision outside of the ground state isotope shift uncertainty to discern any trend.

A change in sign from positive to negative relative isomer shift was observed for the $I^\pi = 1/2^-$ states at $N = 74$, with a minima at $N = 80$ before returning to no difference from zero, within error, at the $N = 82$ shell closure. These changes are smaller than 0.016 fm^2, which is less than that gained for the addition of a neutron ($\sim$0.05 fm^2), (for comparison the change for the high-spin state $I^\pi = 23/2^-$ is $\sim$0.06 fm^2).

Figure 7.13b shows this shift in terms of the change in β_2^{total} value from the $I^\pi = 9/2^+$ state, defining $\delta\beta_2^{total} = \beta_2^{total}(9/2^+) - \beta_2^{total}(1/2^-)$.

As the $I^\pi = 1/2^-$ states have no measurable Q_S then their isomer shifts give an indication of collectivity. The deviation in μ values of the $I^\pi = 1/2^-$ states from their Schmidt value also goes towards negative values for increasing neutron number. Although collective effects may contribute to the deviation [63], this cannot easily be connected with the magnitude of the observed isomer shifts. The μ values of the $I^\pi = 1/2^-$ state begin to increase at $N = 78$, while this occurs abruptly at $N = 82$ for the isomer shift. The scale of this effect in terms of a change in radius due to deformation can been seen on an absolute scale in Fig. 7.12. Compared to the large static and dynamic deformations, the change in deformation for the $I^\pi = 1/2^-$ states is small, and no deviation in the μ values of the $I^\pi = 9/2^+$ states was observed (until $N = 82$) accompanying these large changes in deformation.

The change in sign of the isomer shift may be due to surface effects [55, 64], which have a contribution to the odd-even staggering (OES) of charge radii [57, 65, 66], and which also play a role for high-spin quasi-particle isomers [54, 67] (a significant reduction is also seen for the $I^\pi = 21/2^-$ and $I^\pi = 23/2^-$ states [68]). The

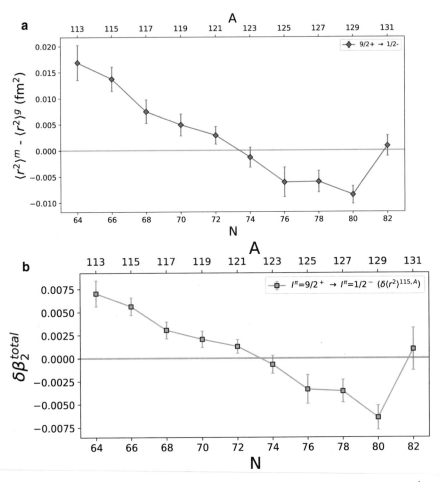

Fig. 7.13 The difference in the **a** charge radii and **b** total deformation parameters, $\delta\beta_2^{total} = \beta_2^{total}(9/2^+) - \beta_2^{total}(1/2^-)$, between the $I^\pi = 1/2^-$ and $I^\pi = 9/2^+$ states, for the odd indium isotopes

abrupt increase in the relative isomer shift at $N = 82$ is due to a reduction in the OES which occurs for the $I^\pi = 9/2^+$ state but not the $I^\pi = 1/2^-$ state. The observed OES is discussed in Sect. 7.3.3.

The shifts in charge radius for the isomer states in the odd indium isotopes are shown in Fig. 7.14, with respect to the ground state, which changes spin from $I^\pi = 1^+$ to $I^\pi = 3^+$ after $N = 74$. Again the charge radii of the excited states decrease in size compared to the ground states for increasing neutron number. It is interesting to note the turning point where the isomers become smaller than the ground state corresponds to the minimum in the excitation energy (minus a neutron) of isomeric states in binding energy. This indicates that the charge radii are highlighting collectivity not discernable in the binding energy (see Sect. 7.3.3 for the OES in binding energy).

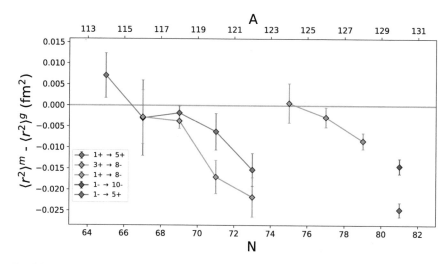

Fig. 7.14 Differences in charge radii of the isomer states compared to the ground states, for the odd indium masses

See the thesis of [68] for a detailed discussion of the high-spin isomer states of ^{127}In, ^{129}In *and* ^{130}In.

7.3.3 The Odd-Even Staggering Effect

A measure of the odd-even staggering (OES) of charge radii can be given by the triple point difference between an isotope charge radii and its neighbouring isotope mass charge radii, $\Delta_r^{(3)}(A)$, as [69]

$$\Delta_r^{(3)}(A) = \frac{1}{2}(\delta r^{115,A+1} - 2\delta r^{115,A} + \delta r^{115,A-1}),\qquad(7.4)$$

where $\delta r^{115,A+1}$, $\delta r^{115,A}$ and $\delta r^{115,A-1}$ are the root-mean-square charge radii.

Figure 7.15 shows the extent of OES observed in the ground and isomeric states of the $^{113-131}$In isotopes measured determined in this work, alongside the OES observed for the neighbouring isotopes tin ($Z = 50$) [40, 70] and cadmium ($Z = 48$) [57].

There are several prominent features in this plot. A reduction in the OES of ground state radii between $N = 70$ and $N = 74$ is now observed, this was not previously seen despite being so close to stability [9], as the ground states are significantly shorter lived than the isomer states in this region (see Fig. 2.3) and so were not accessible. Although the mechanism of odd-even staggering is not fully understood in terms of a quantitative microscopic theory [5, 71–73], it is thought to be due to an orbital blocking effect of the unpaired nucleon and pairing correlations appear to play a significant role [57, 66]. The OES reduction correlates with the jump in μ values

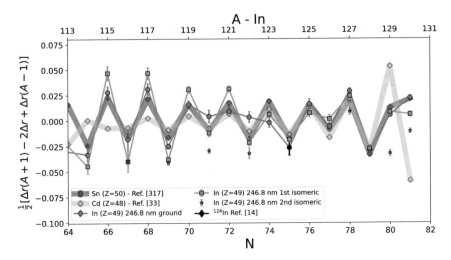

Fig. 7.15 Odd-even staggering of charge radii evaluated using triple difference in charge radii from Eq. 7.4. Compared to literature values for tin ($Z = 50$) [40] and cadmium ($Z = 48$) isotopes [57]

seen in the $I^\pi = 8^-$ ($\pi 1g_{9/2}^{-1} \otimes \nu 1h_{11/2}$) states in the odd isotopes, also attributed to possible neutron-pair correlations.

The Fayans energy density functional (EDF) approach explained the OES of charge radii in terms of neutron pairing density [22, 74, 75]. The Fayans approach has been applied successfully to reproducing the intricate odd-even staggering trend in the calcium isotopes [76–78]. It was found that the reproduction of the OES in the calcium charge radii ($^{40-52}$Ca) was very sensitive to a pairing term which controlled the neutron-pairing density. The changes in mean-squared charge radii were found to be the most important experimental ingredient to constrain the functional [69]. This is shown in Fig. 7.17a. The combination of the Fayan pairing functional term with a Skyrme EDF [69, 79] also was shown to capture the same essential physics in reproducing the calcium radii, although agreement was poorer. Predictions of binding energies were insensitive to both pairing and surface terms in the functionals.

For comparison to the charge radii OES, the triple differences in binding energies of the indium ground states [80, 82] are plotted in Fig. 7.16, using the equivalent equation for the binding energy OES

$$\Delta_E^{(3)}(A) = \frac{1}{2}(B.E._{A+1} - 2B.E._A + B.E._{A-1}) \,, \tag{7.5}$$

with the recent isomer binding energies measured by [81]. The tin, indium and cadmium binding energy OES values show a clear decrease at the $N = 82$ with no obvious discrepancies, other than a reduction in the OES magnitude for the indium isotopes towards the shell closure. Reference [81] observed that the binding energy differentials indicated no signs of deviation from sphericity up to ^{130}In. The single-particle energies indicate a re-ordering of neutrons as they are removed from the

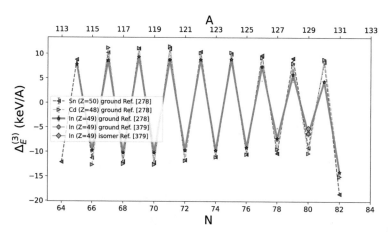

Fig. 7.16 Odd-even staggering of binding energy of indium isotopes evaluated using triple difference in binding energy from Eq. 7.5, compared to cadmium and tin isotopes, as well as indium isomers. Binding energy data from [80] (AME2016), and [81]

$N = 82$ shell however, and the derived empirical proton-neutron correlation energies [83] indicate that the valance proton-neutron interaction is playing a role [81]. The explanation for the reduction in the OES of the ground state charge radii may then be due to pairing effects which do not significantly alter their binding energies and so were not seen as clearly in the binding energy OES as compared to the charge radii OES.

The same Fayan's functionals constrained by the calcium radii were recently applied to make predictions of the cadmium charge radii [57]. The results compared to experimental values are shown in Fig. 7.17b. Although the OES effect in the charge radii are significantly more pronounced in these calculations compared the experimental values [84], the trend is encouraging. The calculations predict that the OES will disappear at $N = 83$ corresponding to the shell closure at $N = 82$, before returning for higher neutron number. This disappearance is seen in the OES of charge radii of indium in this work (Fig. 7.15) for the ground states at $N = 81$, one neutron number earlier than predicted for cadmium, while the OES remains undisturbed from $N = 64$–82 for the isomeric states. Measurements are not yet available to compare to this prediction for cadmium (requires measurement at $N = 83$). The same trend as observed in indium ground states is also found in the tin ($Z = 50$) isotopes [40], shown in Fig. 7.15, again one neutron number earlier, indicating that this disappearance of OES may be due the proximity to the $Z = 50$ shell closure. Measurements of antimony ($Z = 51$) isotope and isomer shifts in the proximity of $Z = 50$ would help understand how universal these phenomenon are and their relationship to the hole or particle nature away from the shell closure. It is interesting to compare the increase in magnitude of the OES in the cadmium isotopes to those in indium. While the first increase in OES staggering is observed for all tin, indium and cadmium at $N = 79$, the disappearance of OES is not seen in cadmium at $N = 81$.

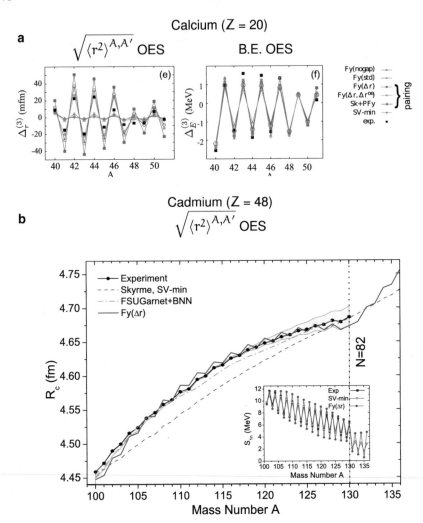

Fig. 7.17 **a** Comparison showing the highly sensitive charge-radii OES for the Fayans functionals with DFT developed by [69], where a the 'pairing gradient functional' was found to be critical to produce OES in calcium radii. Figure modified from [69]. **b** Charge radii calculations of cadmium performed by [57] using the same calcium radii tuned functional as [69]. Figure modified from [57]. © 2020 American Physical Society (Reprinted from Reinhard, P.-G. et al. Toward a global description of nuclear charge radii: Exploring the Fayans energy density functional. PRC 95, 064328 (2017), and Hammen, M. et al. From Calcium to Cadmium: Testing the Pairing Functional through Charge Radii Measurements of Cd100-130. PRL 121, 102501 (2018). Copyright 2020 by the American Physical Society.)

This suggests that the increase in OES at $N = 79$ is due to a separate phenomenon to the disappearance of OES at $N = 81$.

The ground and isomeric states in indium are displaying intricate features in the OES, aspects of which are also seen in the neighbouring cadmium and tin isotopes, but were not observed in the binding energy OES shown in Fig. 7.16. This gives credence to these observed features in the neighbouring isotopes and suggests that reproduction of these trends with developments from nuclear theory should give a valuable insight into the nuclear structure in this area, for which neutron pairing correlations appear to play an important role.

References

1. Schmidt T (1937) Zeitschrift fur Physik 106(5–6):358. https://doi.org/10.1007/BF01338744. ISSN1434-6001
2. Blin-Stoyle RJ (1956) Rev Mod Phys 28(1):75. https://doi.org/10.1103/RevModPhys.28.75. ISSN0034-6861
3. Pastore S et al (2013) Phys Rev C - Nucl Phys 87(3):1. https://doi.org/10.1103/PhysRevC.87. 035503.ISSN05562813
4. Arima A et al (1972) Phys Lett B 41(4):435. https://www.sciencedirect.com/science/article/pii/0370269372906685. ISSN 0370-2693
5. Talmi I (1984) Nucl Phys A 423(2):189. https://www.sciencedirect.com/science/article/pii/0375947484905876. ISSN 0375-9474
6. Stone NJ et al (1997) Phys Rev Lett 78(5):820. https://doi.org/10.1103/PhysRevLett.78.820. ISSN0031-9007
7. Rejmund M et al (2016) Phys Lett B 753:86. http://linkinghub.elsevier.com/retrieve/pii/S0370269315009417. ISSN 03702693
8. Takikawa N et al (2017) Fundamentals of nuclear physics. Springer, Berlin. https://books.google.ch/books?id=DpvlDQAAQBAJ&dq=configuration+mixing+core+polarisation&source=gbs_navlinks_s
9. Eberz J et al (1987) Nucl Phys A 464(1):9. http://linkinghub.elsevier.com/retrieve/pii/0375947487904192. ISSN 03759474
10. Arima A et al (1954) Prog Theor Phys 12(5):623. https://doi.org/10.1143/PTP.12.623. ISSN0033-068X
11. Sheng C et al (2005) J Phys: Conf Ser. https://doi.org/10.1088/1742-6596/20/1/006/pdf
12. Berryman JS et al (2009) Phys Rev C 79(6):064305. https://doi.org/10.1103/PhysRevC.79. 064305.ISSN0556-2813
13. Parzuchowski NM et al (2017) Phys Rev C 96(3):034324. ISSN 24699993
14. Hagen G et al (2016) Physica Scripta 91(063006):1. arXiv:1601.08203. ISSN 14024896
15. Sun ZH et al (2018) Phys Rev C 98(5):054320. https://doi.org/10.1103/PhysRevC.98.054320. ISSN2469-9985
16. Morris TD et al (2018) Phys Rev Lett 120(15):152503. https://doi.org/10.1103/PhysRevLett. 120.152503.ISSN0031-9007
17. Stone N (2005) At Data Nucl Data Tables 90(1):75. http://linkinghub.elsevier.com/retrieve/pii/S0092640X05000239. ISSN 0092640X
18. Heyde K et al (1980) Phys Rev C 22(3):1267. https://doi.org/10.1103/PhysRevC.22.1267. ISSN0556-2813
19. Heyde K (1988) Hyperfine Interact 43(1–4):15. https://doi.org/10.1007/BF02398284. ISSN0304-3843
20. Heyde K et al (1978) Phys Rev C 17(3):1219. https://doi.org/10.1103/PhysRevC.17.1219. ISSN0556-2813
21. Saperstein EE et al (2013) EPL 11:42001. www.epljournal.org

22. Fayans S et al (2000) Nucl Phys A 676(1–4):49. https://www.sciencedirect.com/science/article/pii/S0375947400001925?via%3Dihub. ISSN 0375-9474
23. Neugart R et al (1985) Phys Rev Lett 55(15):1559. https://doi.org/10.1103/PhysRevLett.55.1559.ISSN0031-9007
24. Barzakh AE et al (2013) Phys Rev C 88(2):024315. https://doi.org/10.1103/PhysRevC.88.024315.ISSN0556-2813
25. Bieroń J et al (2001) Phys Rev Lett 87(13):133003. https://doi.org/10.1103/PhysRevLett.87.133003.ISSN0031-9007
26. Hyuga H et al (1980) Nucl Phys A 336(3):363. https://www.sciencedirect.com/science/article/pii/037594748090216X. ISSN 0375-9474
27. Arima AHH et al (1973) J Phys Soc Jpn Suppl 34(205):589
28. Flowers B (1952) Lond Edinb Dublin Philos Mag J Sci 43(347):1330. https://doi.org/10.1080/14786441208520265.ISSN1941-5982
29. Blin-Stoyle RJ (1953) Proc Phys Soc. Sect A 66(12):1158. http://stacks.iop.org/0370-1298/66/i=12/a=312?key=crossref.ac4b2b16288a04160c5fa80bec314e9d. ISSN 0370-1298
30. Byrne J (2013) Neutrons, nuclei, and matter: an exploration of the physics of slow neutrons. Courier Corporation. https://books.google.co.uk/books?id=mTjyAAAAQBAJ&pg=PA188&lpg=PA188&dq=p1/2+magnetic+momentmixing&source=bl&ots=_Vu4ueGiW_&sig=QqHJHEtSrU4daAJjqu5oeRbbqVA&hl=en&sa=X&ved=0ahUKEwjkwtrT6OvVAhUKCsAKHYZ3BwUQ6AEINDAC#v=onepage&q=p1%2F2magneticmomentmixing&
31. Saperstein EE et al (2017) J Phys G: Nucl Part Phys 44(6). ISSN 13616471
32. Orce JN et al (2007) Phys Rev C 76(2):021302. https://doi.org/10.1103/PhysRevC.76.021302.ISSN0556-2813
33. Morrison I et al (1980) Nucl Phys A 350(1–2):89. https://www.sciencedirect.com/science/article/pii/0375947480903905?via%3Dihub. ISSN 0375-9474
34. Allmond JM et al (2015) Phys Rev C 92(4):041303. https://doi.org/10.1103/PhysRevC.92.041303.ISSN0556-2813
35. Jungclaus A et al (2011) Phys Lett B 695(1–4):110. https://www.sciencedirect.com/science/article/pii/S0370269310012864?via%3Dihub. ISSN 0370-2693
36. Togashi T et al (2018) Phys Rev Lett 121(6):062501. https://doi.org/10.1103/PhysRevLett.121.062501.ISSN0031-9007
37. Morales IO et al (2011) Phys Lett B 703(5):606. https://www.sciencedirect.com/science/article/pii/S0370269311009683. ISSN 0370-2693
38. Heyde KLG (1990) The nuclear shell model. Springer series in nuclear and particle physics. Springer, Berlin. https://doi.org/10.1007/978-3-642-97203-4, https://books.google.ch/books?id=aBz4CAAAQBAJ&dq=heyde+nuclear+shell+model&source=gbs_navlinks_s
39. Talmi I (1971) Nucl Phys A 172(1):1. https://www.sciencedirect.com/science/article/pii/0375947471901126?via%3Dihub. ISSN 0375-9474
40. Le Blanc FL et al (2005) Phys Rev C - Nucl Phys 72(3):1. ISSN 05562813
41. Fogelberg B et al (1985) Phys Rev C 31(3):1026. https://doi.org/10.1103/PhysRevC.31.1026.ISSN0556-2813
42. Ionescu-Bujor M et al (1976) Nucl Phys A 272(1):1. https://www.sciencedirect.com/science/article/pii/0375947476903146. ISSN 0375-9474
43. Yordanov DT et al (2013) Phys Rev Lett 110(19):192501. https://doi.org/10.1103/PhysRevLett.110.192501.ISSN0031-9007
44. Stone N (2016) At Data Nucl Data Tables 111–112:1. https://www.sciencedirect.com/science/article/pii/S0092640X16000024. ISSN 0092-640X
45. Scherillo A et al (2005) AIP Conf Proc 798:145. https://doi.org/10.1063/1.2137240.ISSN0094243X
46. Fogelberg B et al (1979) Nucl Phys A 323(2–3):205. http://linkinghub.elsevier.com/retrieve/pii/0375947479901088. ISSN 0375-9474
47. Goeppert-Mayer M et al (2013) Elementary theory of nuclear shell structure. Literary Licensing, LLC. https://books.google.fr/books/about/Elementary_Theory_of_Nuclear_Shell_Struc.html?id=7Mq8mQEACAAJ&source=kp_cover&redir_esc=y

48. Errico LA et al (2006) Phys Rev B 73(11):115125. https://doi.org/10.1103/PhysRevB.73.115125.ISSN1098-0121
49. Brewer WD et al (1978) Hyperfine Interact 5:576. https://doi.org/10.1007/BF01021893.pdf
50. Sternheimer R (1950) Phys Rev 80:102
51. Grupp H et al (1982) Nucl Phys A 386(1):56. https://www.sciencedirect.com/science/article/abs/pii/0375947482904018. ISSN 0375-9474
52. Winnacker A et al (1971) Zeitschrift für Physik A Hadron Nucl 244(4):289. https://doi.org/10.1007/BF01396789.ISSN0939-7922
53. Casserberg BR et al (1968) Puc-937-321. Princeton University
54. Bissell ML et al (2007) Phys Lett Sect B: Nucl Elem Part High-Energy Phys 645(4):330. ISSN 03702693
55. Gangrsky YP (2008) Phys At Nucl 71(7):1168
56. Fricke G et al (2004) 49-In indium. Springer, Berlin. http://materials.springer.com/lb/docs/sm_lbs_978-3-540-45555-4_51, http://materials.springer.com/lb/docs/sm%7B_%7Dlbs%7B_%7D978-3-540-45555-4%7B_%7D51
57. Hammen M et al (2018) Phys Rev Lett 121(10):102501. https://doi.org/10.1103/PhysRevLett.121.102501.ISSN0031-9007
58. Angeli I et al (2013) At Data Nucl Data Tables 99(1):69. https://www.sciencedirect.com/science/article/pii/S0092640X12000265. ISSN 0092-640X
59. Berdichevsky D et al (1985) Z Phys A-At Nucl 322(1):141. https://doi.org/10.1007/BF01412027.ISSN0340-2193
60. Myers WD et al (1983) Nucl Phys A 410(1):61. https://www.sciencedirect.com/science/article/pii/0375947483904013. ISSN 0375-9474
61. Wesolowski E (1985) J Phys G: Nucl Phys 11(8):909. http://stacks.iop.org/0305-4616/11/i=8/a=008?key=crossref.50bd8568f651d17bb5647b0e21f1f90e. ISSN 0305-4616
62. Gangrskii YP et al (1987) J Exp Theor Phys 94:9. http://jetp.ac.ru/cgi-bin/dn/e_067_06_1089.pdf
63. Ichikawa Y et al (2019) Nat Phys. https://doi.org/10.1038/s41567-018-0410-7
64. Gangrsky Y (2006) Hyperfine Interact 171:203
65. Auciello O et al (1993) J Vac Sci Technol A: Vac Surf Films 11(1):267. https://doi.org/10.1116/1.578715.ISSN0734-2101
66. Zawischa D (1985) Phys Lett B 155(5–6):309. https://www.sciencedirect.com/science/article/pii/037026938591576X. ISSN 0370-2693
67. Helmer R et al (1968) Nucl Phys A 114(3):649. https://www.sciencedirect.com/science/article/pii/0375947468902947. ISSN 0375-9474
68. Binnersley C (2019) PhD, The University of Manchester
69. Reinhard P-G et al (2017) Phys Rev C 95(6):064328. https://doi.org/10.1103/PhysRevC.95.064328.ISSN2469-9985
70. Gorges C et al (2019) Phys Rev Lett 122(19):192502. ISSN 10797114
71. Bally B et al (2014) Phys Rev Lett 113(16):162501. https://doi.org/10.1103/PhysRevLett.113.162501.ISSN0031-9007
72. Pototzky KJ et al (2010) Eur Phys J A 46(2):299. https://doi.org/10.1140/epja/i2010-11045-6.ISSN1434-6001
73. Schunck N et al (2010) Phys Rev C 81(2):024316. https://doi.org/10.1103/PhysRevC.81.024316.ISSN0556-2813
74. Fayans S et al (1994) Phys Lett B 338(1):1. https://www.sciencedirect.com/science/article/pii/037026939491334X. ISSN 0370-2693
75. Fayans S et al (1996) Phys Lett B 383(1):19. https://www.sciencedirect.com/science/article/pii/0370269396007162. ISSN 0370-2693
76. Garcia Ruiz RF et al (2016) Nat Phys 12(6):594. https://doi.org/10.1038/nphys3645.ISSN1745-2473
77. Minamisono K et al (2016) Phys Rev Lett 117(25):252501. https://doi.org/10.1103/PhysRevLett.117.252501.ISSN0031-9007

78. Rossi DM et al (2015) Phys Rev C 92(1):014305. https://doi.org/10.1103/PhysRevC.92.014305.ISSN0556-2813
79. Fayans SA (1998) J Exp Theor Phys Lett 68(3):169. https://doi.org/10.1134/1.567841.ISSN0021-3640
80. Wang M et al (2017) Chin Phys C 41(3):30003. http://stacks.iop.org/1674-1137/41/i=3/a=030003?key=crossref.cbf64c2ee5251915f252c1f31e5f8e05. ISSN 1674-1137
81. Babcock C et al (2018) Phys Rev C 97(2):1. ISSN 24699993
82. Spanier L et al (1987) Nucl Phys A 474(2):359. https://www.sciencedirect.com/science/article/pii/037594748790621X?via%3Dihub. ISSN 0375-9474
83. Zhang J-Y et al (1989) Phys Lett B 227(1):1. https://www.sciencedirect.com/science/article/pii/0370269389912732?via%3Dihub. ISSN 0370-2693
84. Yordanov DT et al (2016) Phys Rev Lett 116(3):1. ISSN 10797114

Chapter 8
Conclusion

This thesis presented laser spectroscopy measurements of the electromagnetic properties of the long-lived nuclear ground and isomeric states (>200 ms) of neutron-rich indium isotopes $^{113-131}$In, of which $^{128-131}$In were completely unexplored by laser spectroscopy. The properties of many states in $^{113-128}$In were also previously unmeasured, including the shorter lived ground states. This allowed extraction of 17 new magnetic dipole moment μ values, 15 new quadrupole moments Q_S values, 7 new changes in mean-squared charge radii $\delta \langle r^2 \rangle_g$ values to ground states with sufficient isomer shift precision for 25 new $\delta \langle r^2 \rangle_m$ values for the isomers. In addition the nuclear spins, I, of 11 of these states were confirmed unambiguously. The remaining 26 higher-spin states had values consistent with literature spin values using multiple analysis methods. These measurements were guided and assisted by a series of technical and theoretical developments, which are outlined below along with the findings from these new measurements of the indium isotopes.

The development of a laser ablation ion source setup for offline measurements with beams of the naturally occurring ^{113,115}In isotopes allowed for multi-step laser ionization scheme testing. It was determined that the 246.8-nm (5p ^{2}P$_{3/2}$ $\rightarrow$ 9s ^{2}S$_{1/2}$) and 246.0-nm (5p ^{2}P$_{1/2}$ $\rightarrow$ 8s ^{2}S$_{1/2}$) atomic transitions were the most appropriate for sensitivity to the nuclear electromagnetic observables and their efficient measurement. Simulations were made of the available atomic population following charge exchange of indium ions incident on vapours of potassium and sodium at 20 and 40 keV. The relative populations of the 5p ^{2}P$_{1/2}$ and 5p ^{2}P$_{3/2}$ states were then measured at 20 keV with the ablation source, validating the charge-exchange calculations and their extension to the 40 keV beam energy later used to measure the neutron-rich indium isotopes. The calculations were also found to well reproduce other available relative population measurements and so the calculations were extended to $1 \leq Z \leq 89$, at beam energies of 5 and 40 keV, to assist in the planning of other future collinear laser spectroscopy experiments.

A. R. Vernon, *Collinear Resonance Ionization Spectroscopy of Neutron-Rich Indium Isotopes*, Springer Theses, https://doi.org/10.1007/978-3-030-54189-7_8

The combination of a laser ablation ion source in tandem with collinear resonance ionization spectroscopy was found to be very effective, allowing determination of atomic parameters of potentially any element or molecule which can be contained within a solid matrix. Linewidths of 60(10) MHz were achieved when using low electric field gradients for ion extraction and lineshape corrections using ion time-of-flight information. The measurements additionally give insight into the time dynamics of the laser-ablation plasma itself, a potential area for future studies. The previously unmeasured magnetic dipole hyperfine structure constants A_{hf} of the upper states of these two transitions, $8s\,^2S_{1/2}$ and $9s\,^2S_{1/2}$, provided valuable input to benchmarking of CCSD(T) calculations (performed by [1]). The calculations were used to determine the electronic electric field gradient $\left(\frac{\partial^2 V}{\partial z^2}\right)e = {}^{B_{hf}}\!/Q_S$, improving the accuracy of the nuclear electric quadrupole moment values, Q_S, for the neutron-rich indium isotope measurements.

The isotope shift values for the neutron-rich indium isotopes were then used to test the field shift and mass shift factors of the CCSD(T) calculations. A developed 'analytic response' approach was found to have good agreement with the experimentally determined factors. These atomic factors were used to allow the first isotone independent determination of the $\delta\langle r^2\rangle$ for the indium isotopes. The $\langle r^2\rangle$ values of the $I^\pi = \frac{1}{2}^-$ states for the isomer were observed to be smaller than the ground $I^\pi = \frac{9}{2}^+$ states for $N \geq 67$ in the odd-mass indium isotopes. The effect in terms of deformation was determined to be small, but may be connected to surface effects [2, 3], which were further explored in the odd-even staggering (OES) values of $\delta\langle r^2\rangle$. A reduction in the OES was observed for the ground but not the isomeric states between $N = 70$ and $N = 74$, and then again at $N = 81$, one neutron number earlier than expected for the neutron shell closure at $N = 82$. The same disappearance at $N = 81$ was previously seen in the tin isotopes [4] but not the cadmium isotopes [5], neutron-pair correlations are expected to play an important role [5, 6]. The charge radii of the $I = \frac{9}{2}^+$ states indicate a large dynamic collective contribution at the mid-shell on top of the increasing static contribution determined from the Q_S values, residual static and dynamic collective contributions were found to remain at ^{131}In.

With a few exceptions, the measured μ and Q_S trends for the odd-mass indium isotopes could be reproduced by the additivity rule using the corresponding neutron and proton single-particle values from odd-mass tin (or cadmium) and odd-mass indium isotopes. The moments for the $I = 8^-$ states increase abruptly compared to the predictions, a sudden drop in Q_S accompanied with a increase in μ has now been measured at ^{128}In. Along with the changes in OES observed in $\langle r^2\rangle$, these observations may be explained by neutron pairing correlations in this region of the nuclear chart and the suspected change in structure of the underlying tin core [7].

In the odd-mass In isotopes the first significant change in the μ of the $I = \frac{9}{2}^+$ states was observed at the $N = 82$ shell closure (^{131}In), out of a chain of at least 24 isotopes ($^{105-129}$In) with a remarkably constant μ. This feature was previously attributed to a very small collective or configuration mixing contribution to μ allowing the single-particle behaviour to persist. At the same time it was found that after $N = 76$ the μ values of the accompanying $I = \frac{1}{2}^-$ isomer states increase and then abruptly cross the Schmidt line at $N = 82$. The mechanism behind the change in the $I = 1/2^-$ and $I = 9/2^+$ μ values is not yet clear, but the deviation of the μ values for the $I = 1/2^-$ states supports an explanation in terms of higher-order configuration mixing or meson-exchange currents.

References

1. Sahoo BK (2018) Private communication. CCSD(T) calculations for indium including higher order relativistic effects. Breit and QED effects
2. Auciello O et al (1993) J Vac Sci Technol A: Vac Surf Films 11(1):267. https://doi.org/10.1116/1.578715. ISSN 0734-2101
3. Bissell ML et al (2007) Phys Lett Sect B: Nucl Elem Part High-Energy Phys 645(4):330. ISSN 03702693
4. Le Blanc FL et al (2005) Phys Rev C - Nucl Phys 72(3):1. ISSN 05562813
5. Hammen M et al (2018) Phys Rev Lett 121(10):102501. https://doi.org/10.1103/PhysRevLett.121.102501. ISSN 0031-9007
6. Reinhard P-G et al (2017) Phys Rev C 95(6):064328. https://doi.org/10.1103/PhysRevC.95.064328. ISSN 2469-9985
7. Togashi T et al (2018) Phys Rev Lett 121(6):062501. https://doi.org/10.1103/PhysRevLett.121.062501. ISSN 0031-9007

Appendix A
Additional Formulae

A.1 Wigner 6-j Coefficients

The Wigner 6-j coefficients which arise from the coupling of three angular momenta, were computed by using the Racah formula [1] as

$$\begin{Bmatrix} j_1 & j_2 & j_3 \\ J_1 & J_2 & J_3 \end{Bmatrix} = \sqrt{\Delta(j_1 j_2 j_3)\Delta(j_1 J_2 J_3)\Delta(J_1 j_2 J_3)\Delta(J_1 J_2 j_3)} \times \sum_t \frac{(-1)^t (t+1)!}{f(t)},$$

(A.1)

summed over all integers t if the factorials in

$$\begin{aligned} f(t) = {}& (t - j_1 - j_2 - j_3)! \, (t - j_1 - J_2 - J_3)! \, (t - J_1 - j_2 - J_3)! \\ & (t - J_1 - J_2 - j_3)! \, (j_1 + j_2 + J_2 + J_2 - t)! \, (j_2 + j_3 + J_2 + J_3 - t)! \\ & (j_3 + j_1 + J_3 + J_1 - t)! \,, \end{aligned}$$

all have non-negative values. The $\Delta(abc)$ are the triangle coefficients [2] given by

$$\Delta(abc) \equiv \frac{(a+b-c)!(a-b+c)!(-a+b+c)!}{(a+b+c+1)!}.$$

(A.2)

A.2 Wigner 3-j Coefficients

The Wigner 3-j are a specific case of the 6-j symbols which can also be computed by a Racah relation [1] as

© The Editor(s) (if applicable) and The Author(s), under exclusive license
to Springer Nature Switzerland AG 2020
A. R. Vernon, *Collinear Resonance Ionization Spectroscopy
of Neutron-Rich Indium Isotopes*, Springer Theses,
https://doi.org/10.1007/978-3-030-54189-7

$$\begin{pmatrix} a & b & c \\ \alpha & \beta & \gamma \end{pmatrix} = (-1)^{a-b-\gamma} \tag{A.3}$$

$$\times \sqrt{\Delta(abc)} \sqrt{(a+\alpha)!(a-\alpha)!(b+\beta)!(b-\beta)!(c+\gamma)!(c-\gamma)!} \tag{A.4}$$

$$\times \sum_t \frac{(-1)^t}{x}, \tag{A.5}$$

where the condition for the sum over t is the same and x is given by

$$x = t!(c-b+t+\alpha)!(c-a+t-\beta)!(a+b-c-t)! \tag{A.6}$$
$$\times (a-t-\alpha)!(b-t+\beta)!. \tag{A.7}$$

A.3   Doppler Shift

To reverse the Doppler effect in the frequency of the transitions $\nu_{Dop.}$ measured, into the rest frame frequencies, ν_{rest}, the Doppler shift formula

$$\nu_{rest} = \nu_{Dop.} \sqrt{\frac{1 \pm \beta}{1 \mp \beta}}, \tag{A.8}$$

was used. Where the $\pm$ sign corresponds to a collinear geometry ($+$) or an anti-collinear geometry ($-$), $\beta = v/c$, the speed of light c.

A.4   Measurement Time

Statistical significance, σ, and measurement time, t, relationship for a signal rate, S and background rate, B calculated by [3]

$$\sigma = \frac{St - \sqrt{St}}{\sqrt{Bt}}. \tag{A.9}$$

A.5   Propagation of Uncertainties

For uncorrelated independent variables x, y, z, ...related by a function F, with uncertainties of the variables σ_x, σ_y, σ_z, ..., the uncertainty of the function σ_F can be propagated by [3]

$$\sigma_F = \sqrt{\left(\frac{\partial F(x, y, z, \ldots)}{\partial x}\sigma_x\right)^2 + \left(\frac{\partial F(x, y, z, \ldots)}{\partial y}\sigma_y\right)^2 + \left(\frac{\partial F(x, y, z, \ldots)}{\partial z}\sigma_z\right)^2 + \ldots} \tag{A.10}$$

A.6 Quadrupole Explicit Formula

$$Q_0 = \frac{3}{\sqrt{5\pi}} Z e R^2 \beta_2 \left(1 + \pi^2 \left(\frac{a}{R}\right)^2 + \frac{2}{7}\sqrt{\frac{5}{\pi}}\beta_2\right), \tag{A.11}$$

where a is an empirical correction for the surface diffuseness and R is the radius of an homogeneous sphere [4], which when related to mass A of the nucleus (usually by $R = 1.2A^{1/3}$ fm), can allow for a measure of the *static* deformation β_2 of the nucleus.

A.7 Muonic Atom Calibration

The nuclear radius parameter $\Lambda_{\mu e}^{113,115} = 0.157(11)$ fm^2 was determined from muonic atom transition [5] measurements of ^{113,115}In. This parameter is related to the change in mean-squared charge radii $\delta \langle r^2 \rangle_{o\mu e}^{113,115}$ by

$$\Lambda_{\mu e}^{113,115} = \delta \langle r^2 \rangle_{o\mu e}^{113,115} + HM, \tag{A.12}$$

where HM are the higher radial moments. From the Seltzer moments of indium [8] this contribution should be less than 0.06%, this can be included as an error to give value of $\delta \langle r^2 \rangle_{o\mu e}^{113,115} = 0.157(11)$ fm^2.

A.8 ISCOOL Voltage Calibration

See Fig. A.1.

Fig. A.1 A comparison of measured voltages between a higher precision reference high voltage divider [7] and the divider installed at ISCOOL during these experiments. This gave the calibration function $V_{correct} = 0.99877(1) \times V_{measured} + 0.027(22)$

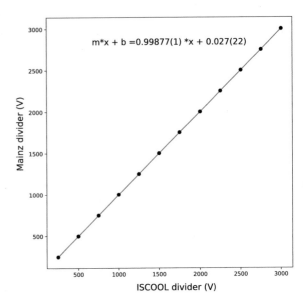

A.9 ISCOOL Correlation Plots

See Figs. A.2 and A.3.

Fig. A.2 Correlation plot of the ^{115}In hyperfine spectra centroids for the 246.0-nm transition versus the recorded ISCOOL extraction voltage. r is the correlation coefficient

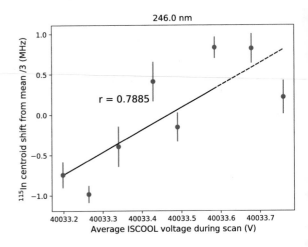

Fig. A.3 Correlation plot of the ^{115}In hyperfine spectra centroids for the 246.8-nm transition versus the recorded ISCOOL extraction voltage. r is the correlation coefficient

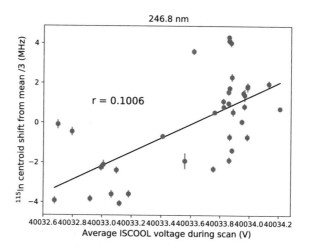

A.10 Additivity Rule Values Using Cadmium Coupling

See Figs. A.4 and A.5.

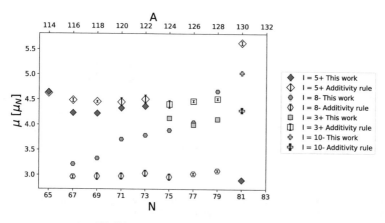

Fig. A.4 The determined $^{113-131}$In magnetic dipole moment values compared to additivity rule values using the $\mu(J_n)$ values from the cadmium isotopes [8]

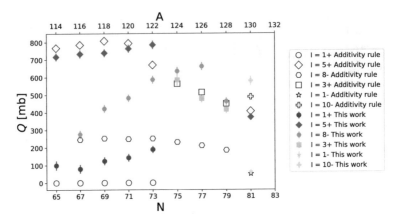

Fig. A.5 The determined $^{113-131}$In electric quadrupole moment values compared to additivity rule values using the $Q(J_n)$ values from the cadmium isotopes [8]

A.11 ^{130}In F Ratio Spin Assignment

See Fig. A.6.

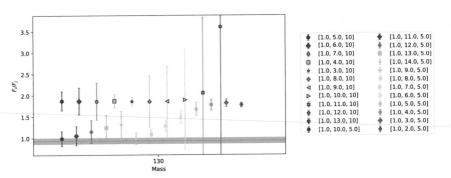

Fig. A.6 Spin assignment of the structures in ^{130}In from the F_i/F_j ratios. The labels indicate the fits for: [$I = 1$ structure, centre-most structure, other structure]

A.12 ^{131}In A Ratio Spin Assignment

See Fig. A.7.

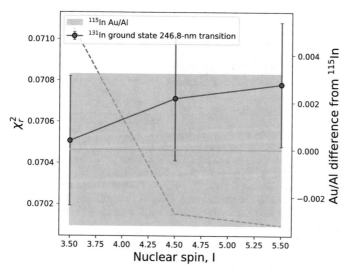

Fig. A.7 The A_u / A_l ratios from fitting the higher spin state in ^{131}In with $I = 7/2$, $I = 9/2$ or $I = 11/2$. Showing the insensitivity to spin of the $5p\ ^2P_{3/2} \rightarrow 9s\ ^2S_{1/2}$ (246.8-nm) transition

A.13 Sidepeak Amplitude Fitting

See Fig. A.8.

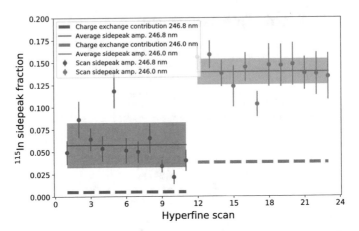

Fig. A.8 Fitted amplitudes of the sidepeak due to indirect population of the 5p ^{2}P$_{1/2}$ and 5p ^{2}P$_{3/2}$ states for the for the 5p ^{2}P$_{1/2} \to$ 8s ^{2}S$_{1/2}$ (246.0-nm) and 5p ^{2}P$_{3/2} \to$ 9s ^{2}S$_{1/2}$ (246.8-nm) transitions respectively. Only spectra with $0.8 < \chi_r < 1.2$ were included in the average. Error on the averages are indicated by the shaded area. Predicted contribution from intermediate charge exchange populations are indicated by the dashed line. The remaining contribution is likely due to collisional excitation

A.14   Free Proton and Neutron Magnetic Moment Comparison

See Fig. A.9.

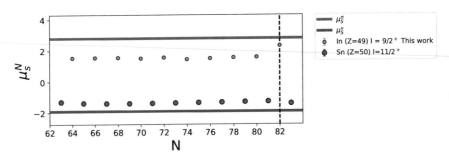

Fig. A.9 Magnetic moments μ of the free proton and neutron, compared to the μ values of the proton hole and neutron particle ground states in the indium and tin isotopes [9]

A.15   Cross-Experiment King Plots

See Figs. A.10 and A.11.

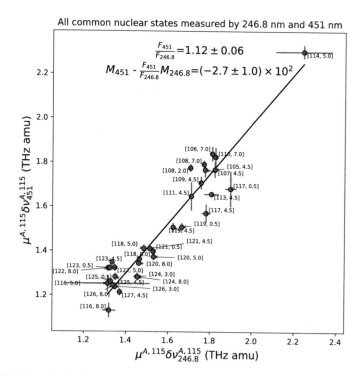

Fig. A.10 King plot of all nuclear states measured with both 246.8-nm ($5p\,^2P_{3/2} \rightarrow 9s\,^2S_{1/2}$) and 451 nm ($5p\,^2P_{3/2} \rightarrow 6s\,^2S_{1/2}$) transitions

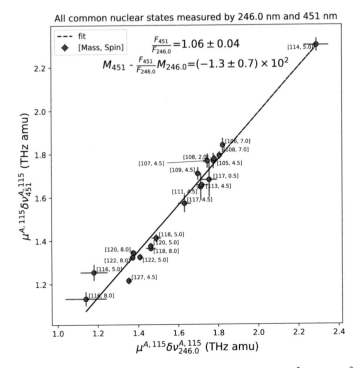

Fig. A.11 King plot of all nuclear states measured with both 246.0 nm (5p $^2P_{1/2} \to$ 8s $^2S_{1/2}$) and 451 nm (5p $^2P_{3/2} \to$ 6s $^2S_{1/2}$) transitions

References

1. Messiah A (1999) Quantum mechanics. Dover Publications, Mineola. https://books.google.ch/books/about/Quantum_Mechanics.html?id=mwssSDXzkNcC&redir_esc=y
2. Shore BW et al (1967) Principles of atomic spectra. Wiley, New York
3. Knoll GF (2010) Radiation detection and measurement. Wiley, New York
4. Rowe DJ (2010) The collective rotational model. World Scientific, Singapore. http://www.worldscientific.com/doi/abs/10.1142/9789812790668_0006
5. Jansen J (1989) Diploma thesis, University of Mainz
6. Seltzer EC (1969) Phys Rev 188(4):1916. https://journals.aps.org/pr/abstract/10.1103/PhysRev.188.1916
7. Krieger A et al (2011) Nuclear Instruments and Methods in Physics Research Section A: Accelerators, Spectrometers, Detectors and Associated Equipment 632(1):23. ISSN 0168-9002. https://www.sciencedirect.com/science/article/pii/S0168900210029256?via%3Dihub
8. Yordanov DT, et al (2013) Phys Rev Lett 110(19):192501. ISSN 0031-9007. https://link.aps.org/doi/10.1103/PhysRevLett.110.192501
9. Le Blanc FL et al (2005) Phys Rev C - Nuclear Phys 72(3):1. ISSN 05562813

Appendix B
Charge-Exchange Calculation Result Tables

B.1 Table: Relative Atomic Populations Following Neutralization with K

See Table B.1.

Table B.1 Atomic population distribution simulation results for neutralisation of an ion beam of elements A = 1 − 90 by free K atoms. Atomic data sourced from the NIST atomic database [1]. Only the 5 most populous states are listed in this table. ★—these light elements are outside of the intermediate velocity region at 40 keV. Some elements were not included if insufficient level or transition data was available to provide reliable values for these elements. The columns for 0 cm and 120 cm flight distance give respectively the initial populations after charge exchange at 0 cm and the final populations after a further 120 cm of atom flight at the corresponding beam energy

$A^+ + K \rightarrow A(k) + K^+$

A^+		40 keV				5 keV			
		Initial (0.0 cm)		Final (120.0 cm)		Initial (0.0 cm)		Final (120.0 cm)	
(Z)	#	Level (cm^{-1})	%	Level (cm^{-1})	%	Level (cm^{-1})	%	Level (cm^{-1})	%
1★	1	109610.2232	1.0	0.0	40.9	82258.91911	3.9	0.0	61.0
	2	109606.6628	1.0	82258.9544	1.8	82258.9544	3.9	82258.9544	4.4
	3	109602.8177	1.0	109583.8949	1.0	82259.158	3.9	107440.4385	0.8
	4	109598.6566	1.0	109538.8768	1.0	82259.285	3.9	107440.43933	0.8
	5	109594.1439	1.0	109548.3584	1.0	97492.2112	1.9	107965.04916	0.7

(continued)

© The Editor(s) (if applicable) and The Author(s), under exclusive license
to Springer Nature Switzerland AG 2020
A. R. Vernon, *Collinear Resonance Ionization Spectroscopy
of Neutron-Rich Indium Isotopes*, Springer Theses,
https://doi.org/10.1007/978-3-030-54189-7

Table B.1 (continued)

$A^+ + K \rightarrow A(k) + K^+$

| (Z) | # | 40 keV | | | | 5 keV | | | |
| | | Initial (0.0 cm) | | Final (120.0 cm) | | Initial (0.0 cm) | | Final (120.0 cm) | |
		Level (cm^{-1})	%	Level (cm^{-1})	%	Level (cm^{-1})	%	Level (cm^{-1})	%
2*	1	171134.89695	1.0	159855.97433	52.3	166277.44014	10.6	159855.97433	75.1
	2	169087.83081	1.0	0.0	9.1	159855.97433	9.5	0.0	7.9
	3	169086.8429	1.0	193921.61495	0.7	169086.76647	9.0	166277.44014	4.5
	4	169086.76647	1.0	193921.61772	0.7	169086.8429	9.0	193921.61495	0.3
	5	166277.44014	1.0	193921.62024	0.7	169087.83081	9.0	193921.61772	0.3
3*	1	14903.66	3.5	0.0	51.5	14903.66	27.2	0.0	88.9
	2	14904.0	3.5	42003.3	0.7	14904.0	27.2	42003.3	0.2
	3	0.0	2.8	42298.0	0.7	0.0	17.9	42298.0	0.2
	4	27206.12	2.4	42389.0	0.7	27206.12	2.4	42389.0	0.2
	5	30925.38	1.9	42389.0	0.7	30925.38	1.2	42389.0	0.2
4	1	42565.35	3.5	21978.28	65.9	42565.35	50.8	21978.28	50.8
	2	52080.94	3.0	21978.925	14.8	52080.94	7.7	0.0	29.6
	3	54677.26	2.6	0.0	6.2	54677.26	4.1	21978.925	11.6
	4	56882.43	2.3	21981.27	4.7	21981.27	2.6	21981.27	4.5
	5	58907.45	2.1	56882.43	1.6	21978.925	2.6	56882.43	1.7
5	1	40039.6907	5.3	0.0	39.8	28658.4	26.7	28658.4	26.7
	2	28658.4	5.2	15.287	14.8	28652.07	26.7	28652.07	26.7
	3	28652.07	5.2	28658.4	5.1	28647.43	26.7	28647.43	26.7
	4	28647.43	5.2	28652.07	5.1	40039.6907	10.1	0.0	16.0
	5	47856.809	3.6	28647.43	5.1	47856.809	1.3	15.287	2.0
6	1	60333.43	1.9	16.4	37.0	60333.43	15.2	16.4	51.9
	2	60352.63	1.9	0.0	21.2	60352.63	15.1	0.0	31.1
	3	60393.14	1.9	43.4	13.4	60393.14	15.0	43.4	8.8
	4	61981.82	1.8	10192.63	6.1	61981.82	10.5	10192.63	5.7
	5	64086.92	1.7	33735.2	0.4	64086.92	6.2	33735.2	0.2
7	1	83364.62	1.9	0.0	54.4	83284.07	15.2	0.0	83.8
	2	83317.83	1.9	19224.464	8.3	83317.83	15.1	19224.464	7.0
	3	83284.07	1.9	19233.177	5.1	83364.62	15.1	19233.177	6.8
	4	86137.35	1.9	28838.92	1.1	86137.35	10.0	28838.92	0.1
	5	86220.51	1.9	106868.635	1.0	86220.51	9.9	106868.635	0.1
8	1	76794.978	4.4	73768.2	47.0	73768.2	41.8	73768.2	62.1
	2	73768.2	4.3	0.0	20.2	76794.978	39.1	0.0	34.5
	3	86625.757	3.0	158.265	4.3	86625.757	2.3	158.265	0.8
	4	86627.778	3.0	102865.655	0.9	86627.778	2.2	102865.655	0.1
	5	86631.454	3.0	226.977	0.6	86631.454	2.2	105441.645	0.0
9	1	105056.28	4.1	102405.71	18.1	105056.28	20.6	102405.71	24.6
	2	104731.05	4.1	102680.44	14.6	104731.05	20.1	102680.44	22.6
	3	102840.38	3.9	102840.38	8.5	102840.38	15.0	102840.38	20.0
	4	102680.44	3.8	104731.05	7.7	102680.44	14.4	104731.05	15.0
	5	102405.71	3.8	105056.28	4.4	102405.71	13.5	105056.28	13.7

(continued)

Table B.1 (continued)

$A^+ + K \rightarrow A(k) + K^+$

A^+		40 keV				5 keV			
		Initial (0.0 cm)		Final (120.0 cm)		Initial (0.0 cm)		Final (120.0 cm)	
(Z)	#	Level (cm^{-1})	%	Level (cm^{-1})	%	Level (cm^{-1})	%	Level (cm^{-1})	%
10	1	135888.7173	4.1	0.0	23.5	135888.7173	25.2	0.0	44.6
	2	134818.6405	3.9	134041.84	22.1	134818.6405	18.8	134041.84	32.3
	3	134459.2871	3.9	134818.6405	5.6	134459.2871	16.9	134818.6405	16.0
	4	134041.84	3.8	158601.1152	2.5	134041.84	14.9	158601.1152	0.4
	5	148257.7898	3.2	159379.9935	1.1	148257.7898	3.2	159379.9935	0.2
11	1	0.0	14.0	0.0	71.8	0.0	60.7	0.0	94.5
	2	16956.17025	11.4	37059.54	0.6	16956.17025	12.3	37059.54	0.1
	3	16973.36619	11.4	38400.9	0.5	16973.36619	12.3	38400.9	0.1
	4	25739.999	3.5	38400.904	0.5	25739.999	1.0	38400.904	0.1
	5	29172.837	2.1	38401.147	0.5	29172.837	0.5	38401.147	0.1
12	1	21911.178	8.7	21870.464	26.8	21911.178	28.6	21850.405	32.5
	2	21870.464	8.7	21850.405	24.8	21870.464	28.2	21870.464	31.4
	3	21850.405	8.7	21911.178	18.6	21850.405	28.0	21911.178	31.2
	4	35051.264	7.6	0.0	7.9	35051.264	7.3	0.0	3.2
	5	41197.403	3.7	46403.065	3.0	41197.403	0.9	46403.065	0.3
13	1	25347.756	9.2	0.0	40.1	25347.756	17.5	0.0	49.2
	2	112.061	7.4	112.061	22.9	112.061	17.3	112.061	24.4
	3	0.0	7.3	29020.41	5.6	0.0	16.8	29020.41	5.8
	4	29020.41	5.6	29066.96	5.5	29020.41	5.8	29066.96	5.8
	5	29066.96	5.5	29142.78	5.5	29066.96	5.8	29142.78	5.7
14	1	33326.053	8.0	77.115	29.7	33326.053	68.8	33326.053	52.5
	2	39683.163	5.2	0.0	23.5	39683.163	5.3	77.115	22.5
	3	39760.285	5.2	223.157	7.4	39760.285	5.1	0.0	14.0
	4	39955.053	5.1	6298.85	4.7	39955.053	4.7	223.157	3.1
	5	40991.884	4.5	33326.053	4.6	40991.884	3.2	6298.85	1.8
15	1	55939.421	4.5	0.0	32.3	55939.421	19.0	0.0	75.3
	2	56090.626	4.5	11361.02	6.8	56090.626	17.8	11361.02	9.2
	3	56339.656	4.4	65788.455	2.1	56339.656	15.9	65373.556	0.7
	4	57876.574	3.7	65585.13	2.1	57876.574	8.1	65450.125	0.7
	5	58174.366	3.6	65450.125	2.1	58174.366	7.1	65585.13	0.7
16	1	52623.64	11.0	0.0	39.5	52623.64	65.3	0.0	46.2
	2	55330.811	9.0	52623.64	32.5	55330.811	19.6	52623.64	41.4
	3	63446.065	3.2	63446.065	14.2	63446.065	1.0	396.055	8.9
	4	63457.142	3.2	396.055	6.5	63457.142	1.0	63446.065	2.7
	5	63475.051	3.1	573.64	2.3	63475.051	1.0	573.64	0.3
17	1	71958.363	6.9	0.0	43.9	71958.363	29.3	0.0	71.5
	2	72488.568	6.7	71958.363	12.4	72488.568	24.0	882.3515	20.2
	3	72827.038	6.6	882.3515	12.0	72827.038	20.9	71958.363	6.5
	4	74225.846	6.1	87979.49	1.3	74225.846	11.2	87979.49	0.1
	5	74865.667	5.8	88080.042	1.3	74865.667	8.3	88080.042	0.1

(continued)

Table B.1 (continued)

$A^+ + K \rightarrow A(k) + K^+$

| (Z) | # | 40 keV | | | | | | 5 keV | | | | | |
| | | Initial (0.0 cm) | | Final (120.0 cm) | | | | Initial (0.0 cm) | | Final (120.0 cm) | |
		Level (cm^{-1})	%	Level (cm^{-1})	%			Level (cm^{-1})	%	Level (cm^{-1})	%
18	1	93143.76	9.9	0.0	31.2			93143.76	32.9	0.0	43.6
	2	93750.5978	9.8	93143.76	30.2			93750.5978	28.2	93143.76	39.0
	3	94553.6652	9.5	94553.6652	10.3			94553.6652	21.1	94553.6652	16.2
	4	95399.8276	9.2	112750.153	1.2			95399.8276	14.5	112750.153	0.1
	5	104102.099	3.0	113716.555	0.9			104102.099	0.3	113716.555	0.0
19	1	0.0	31.8	0.0	68.3			0.0	97.2	0.0	99.0
	2	12985.18572	8.4	30617.31	0.4			12985.18572	0.5	30617.31	0.0
	3	13042.89603	8.3	30617.31	0.4			13042.89603	0.5	30617.31	0.0
	4	21026.551	1.9	33910.42	0.2			21026.551	0.1	34359.36	0.0
	5	21534.68	1.8	33910.42	0.2			21534.68	0.1	34510.119	0.0
20	1	15157.901	9.7	15210.063	13.8			15157.901	29.3	15157.901	30.0
	2	15210.063	9.7	15157.901	13.2			15210.063	29.1	15210.063	29.1
	3	15315.943	9.6	15315.943	13.0			15315.943	28.5	15315.943	29.1
	4	20335.36	7.0	20335.36	11.9			20335.36	3.1	20335.36	3.3
	5	20349.26	7.0	20349.26	10.8			20349.26	3.1	20349.26	3.2
21	1	18504.06	2.3	0.0	13.8			18504.06	4.7	0.0	15.8
	2	18515.69	2.3	168.34	11.4			18515.69	4.7	168.34	14.7
	3	18571.41	2.3	11519.99	5.4			18571.41	4.7	16210.85	5.5
	4	18711.02	2.3	11557.69	4.3			17307.08	4.6	16026.62	5.1
	5	18855.74	2.3	11610.28	4.1			17255.07	4.6	16009.77	5.1
22	1	20795.603	1.0	0.0	9.4			20126.062	3.1	0.0	9.7
	2	20209.447	1.0	386.874	7.5			20062.886	3.1	386.874	9.0
	3	20126.062	1.0	170.1328	7.3			20209.447	3.1	170.1328	7.7
	4	20062.886	1.0	6598.765	4.5			20006.039	3.1	6842.962	3.9
	5	20006.039	1.0	6556.833	3.7			19937.855	3.1	6598.765	3.6
23	1	20202.47	1.0	2153.21	7.3			19189.33	2.6	2220.11	9.2
	2	19189.33	1.0	0.0	6.2			19145.13	2.6	2153.21	8.7
	3	20606.5	1.0	2220.11	6.2			19078.11	2.6	2424.78	7.9
	4	19145.13	1.0	2112.28	5.6			19026.33	2.6	2311.36	7.9
	5	20687.76	1.0	2311.36	5.4			19023.52	2.6	0.0	6.1
24	1	20517.4222	1.9	0.0	12.7			20517.4222	8.5	0.0	10.7
	2	20519.5515	1.9	7593.1484	5.7			20519.5515	8.5	20517.4222	8.6
	3	20520.9029	1.9	7810.7795	2.3			20520.9029	8.5	20519.5515	8.6
	4	20523.629	1.9	20519.5515	2.3			20523.629	8.5	20520.9029	8.6
	5	20523.8999	1.9	20517.4222	2.2			20523.8999	8.5	20523.8999	8.6
25	1	25287.74	3.0	0.0	18.0			24802.25	7.8	0.0	32.1
	2	25285.43	3.0	18402.46	3.4			24788.05	7.8	25265.74	6.9
	3	25281.04	3.0	17451.52	3.2			24779.32	7.8	25281.04	6.8
	4	25265.74	3.0	17052.29	3.1			25265.74	7.8	25285.43	6.8
	5	24802.25	3.0	23296.67	2.9			25281.04	7.8	25287.74	6.8

(continued)

Table B.1 (continued)

$A^+ + K \rightarrow A(k) + K^+$

| A^+ | | 40 keV | | | | 5 keV | | | |
| | | Initial (0.0 cm) | | Final (120.0 cm) | | Initial (0.0 cm) | | Final (120.0 cm) | |
(Z)	#	Level (cm^{-1})	%	Level (cm^{-1})	%	Level (cm^{-1})	%	Level (cm^{-1})	%
26	1	29056.324	1.1	0.0	12.3	28819.954	3.7	0.0	16.4
	2	29313.008	1.1	415.933	9.0	28604.613	3.7	415.933	14.4
	3	29320.026	1.1	704.007	8.4	29056.324	3.7	704.007	13.6
	4	29356.744	1.1	888.132	6.4	29313.008	3.6	888.132	9.0
	5	29371.812	1.1	7376.764	3.0	29320.026	3.6	29320.026	3.3
27	1	28845.22	1.6	0.0	11.3	28470.51	4.8	0.0	20.9
	2	28777.27	1.6	816.0	8.7	28777.27	4.8	816.0	16.7
	3	29216.37	1.6	1406.84	7.8	28845.22	4.8	1406.84	11.7
	4	29269.73	1.6	1809.33	6.0	28345.86	4.8	1809.33	6.7
	5	29294.52	1.6	3482.82	5.1	29216.37	4.5	26597.64	4.4
28	1	27260.894	2.6	204.787	18.5	26665.887	7.2	0.0	22.9
	2	27414.868	2.6	0.0	15.5	27260.894	6.8	204.787	17.4
	3	26665.887	2.6	879.816	14.7	27414.868	6.6	879.816	9.3
	4	27580.391	2.6	1713.087	7.9	25753.553	6.3	27260.894	7.7
	5	27943.524	2.6	1332.164	5.5	27580.391	6.3	25753.553	7.2
29	1	30535.324	21.6	0.0	58.1	30535.324	50.9	0.0	95.9
	2	30783.697	21.0	11202.618	7.0	30783.697	43.5	11202.618	0.8
	3	39018.69	4.7	39018.69	4.7	39018.69	0.7	39018.69	0.7
	4	40114.01	3.7	13245.443	3.9	13245.443	0.5	13245.443	0.5
	5	13245.443	3.3	40909.16	3.1	40114.01	0.4	40909.16	0.3
30	1	46745.413	22.3	32890.352	19.5	46745.413	41.9	32890.352	26.8
	2	32890.352	14.8	32501.421	18.2	32890.352	18.7	32501.421	22.6
	3	32501.421	13.8	32311.35	17.6	32501.421	15.7	32311.35	20.8
	4	32311.35	13.3	46745.413	9.8	32311.35	14.5	46745.413	20.0
	5	53672.28	5.3	53672.28	7.0	53672.28	2.0	53672.28	2.9
31	1	24788.53	18.5	0.0	36.5	24788.53	22.1	0.0	42.7
	2	826.19	15.9	826.19	21.1	826.19	21.3	826.19	24.8
	3	0.0	13.6	33044.05	2.8	0.0	16.7	33044.05	2.2
	4	33044.05	2.8	33155.07	2.8	33044.05	2.2	33155.07	2.2
	5	33155.07	2.8	37975.768	1.1	33155.07	2.2	37975.768	0.8
32	1	37451.6893	15.7	557.1341	39.1	37451.6893	22.6	557.1341	44.9
	2	37702.3054	14.8	0.0	24.4	37702.3054	20.0	0.0	29.8
	3	39117.9021	10.5	41926.726	3.1	39117.9021	10.9	41926.726	2.3
	4	40020.5604	8.4	1409.9609	3.0	40020.5604	7.8	1409.9609	2.0
	5	16367.3332	7.5	45985.592	2.6	16367.3332	7.7	45985.592	1.8
33	1	50693.8	16.8	50693.8	20.6	50693.8	32.0	50693.8	36.6
	2	51610.2	13.7	51610.2	16.8	51610.2	19.3	51610.2	22.1
	3	52897.9	10.1	52897.9	12.3	52897.9	10.2	52897.9	11.7
	4	53135.6	9.5	55366.4	6.6	53135.6	9.2	53135.6	5.3
	5	54605.3	6.6	53135.6	5.8	54605.3	5.0	55366.4	4.3

(continued)

Table B.1 (continued)

$A^+ + K \rightarrow A(k) + K^+$

| (Z) | # | 40 keV | | | | | | 5 keV | | | | | | |
| | | Initial (0.0 cm) | | Final (120.0 cm) | | Initial (0.0 cm) | | Final (120.0 cm) | | | | | | |
		Level (cm^{-1})	%	Level (cm^{-1})	%	Level (cm^{-1})	%	Level (cm^{-1})	%
35	1	63436.45	22.0	63436.45	26.2	63436.45	60.1	63436.45	63.1
	2	64907.19	17.5	64907.19	20.8	64907.19	21.5	64907.19	22.6
	3	66883.87	11.6	66883.87	13.8	66883.87	6.3	66883.87	6.6
	4	67183.58	10.8	67183.58	6.4	67183.58	5.3	67183.58	2.8
	5	68970.21	7.0	68970.21	4.2	68970.21	2.2	68970.21	1.1
36	1	79971.7417	26.2	79971.7417	34.2	79971.7417	63.1	79971.7417	64.2
	2	80916.768	23.6	80916.768	28.1	80916.768	31.9	80916.768	32.4
	3	85191.6166	10.1	85191.6166	8.1	85191.6166	1.7	85191.6166	1.2
	4	85846.7046	8.6	85846.7046	7.1	85846.7046	1.2	85846.7046	0.9
	5	91168.515	2.1	91168.515	1.7	91168.515	0.1	91168.515	0.1
37	1	0.0	68.6	0.0	74.1	0.0	99.2	0.0	99.3
	2	12578.95	4.3	12816.545	4.2	12578.95	0.1	12816.545	0.1
	3	12816.545	4.0	19355.203	0.8	12816.545	0.1	19355.203	0.0
	4	19355.203	0.8	19355.649	0.8	19355.203	0.0	19355.649	0.0
	5	19355.649	0.8	20132.51	0.7	19355.649	0.0	20132.51	0.0
38	1	14317.507	18.4	14317.507	21.2	14317.507	35.7	14317.507	36.4
	2	14504.334	17.9	14898.545	18.8	14504.334	30.9	14898.545	23.2
	3	14898.545	16.7	14504.334	16.8	14898.545	22.8	14504.334	18.4
	4	18159.04	8.2	18159.04	10.1	18159.04	2.5	0.0	13.5
	5	18218.784	8.1	18218.784	9.6	18218.784	2.5	18159.04	2.7
39	1	15476.533	3.3	0.0	8.0	15221.633	6.4	0.0	6.5
	2	15328.865	3.3	530.351	5.3	15245.803	6.4	15221.633	6.4
	3	15326.741	3.3	15326.741	3.5	15326.741	6.4	15245.803	6.4
	4	15245.803	3.3	14948.994	3.4	15328.865	6.4	15326.741	6.4
	5	15221.633	3.3	15221.633	3.4	14948.994	6.3	15328.865	6.4
42	1	22244.375	1.8	0.0	10.7	22244.375	6.4	22244.375	5.4
	2	22875.942	1.8	10768.332	3.4	21618.63	5.8	21618.63	4.9
	3	21618.63	1.8	20280.968	2.0	22875.942	5.7	22875.942	4.8
	4	23516.476	1.8	20350.507	2.0	21343.204	5.3	0.0	4.7
	5	23534.407	1.8	20130.313	2.0	21153.884	4.8	21343.204	4.5
43	1	22615.94	1.9	0.0	10.5	22503.1	4.9	0.0	19.6
	2	22503.1	1.9	2572.89	2.8	22265.46	4.9	22503.1	5.6
	3	22265.46	1.9	17522.92	2.1	22615.94	4.9	22265.46	5.6
	4	22180.1	1.9	22503.1	2.0	22180.1	4.8	21920.15	5.2
	5	23265.32	1.9	22265.46	2.0	21926.79	4.6	21752.14	5.0
44	1	24927.48	2.1	0.0	6.5	24173.68	7.4	25214.16	10.2
	2	24173.68	2.1	1190.64	5.1	24927.48	6.8	25464.49	9.0
	3	25214.16	2.0	25214.16	2.9	25214.16	6.1	0.0	7.9
	4	25464.49	2.0	25464.49	2.9	23453.47	5.9	26035.56	6.3
	5	25602.6	2.0	26035.56	2.8	23392.6	5.7	25602.6	5.0

(continued)

Table B.1 (continued)

$A^+ + K \rightarrow A(k) + K^+$

(Z)	#	40 keV						5 keV			
		Initial (0.0 cm)		Final (120.0 cm)				Initial (0.0 cm)		Final (120.0 cm)	
		Level (cm^{-1})	%	Level (cm^{-1})	%			Level (cm^{-1})	%	Level (cm^{-1})	%
45	1	25820.8	6.0	0.0	25.8			24686.0	23.3	0.0	26.1
	2	24686.0	6.0	1529.97	14.5			25820.8	22.2	25820.8	20.5
	3	27075.26	5.5	2598.03	13.6			23655.93	13.9	24686.0	18.5
	4	23655.93	5.5	25820.8	4.6			27075.26	10.4	23655.93	11.0
	5	23157.57	5.2	24686.0	3.9			23157.57	9.8	23157.57	7.7
46	1	34068.977	9.5	6564.148	20.6			34068.977	29.8	29711.109	25.5
	2	35041.751	8.4	29711.109	12.6			29711.109	17.3	34068.977	23.5
	3	29711.109	8.2	28213.767	9.6			35041.751	13.6	6564.148	18.1
	4	35451.443	7.8	0.0	7.9			35451.443	9.7	35041.751	12.0
	5	35927.948	7.2	34068.977	7.7			35927.948	6.6	28213.767	8.0
47	1	29552.05741	32.0	0.0	59.8			29552.05741	47.4	0.0	70.5
	2	30242.29835	27.8	30242.29835	27.8			30242.29835	27.4	30242.29835	27.4
	3	30472.66516	26.5	34714.22643	8.6			30472.66516	23.0	34714.22643	1.8
	4	34714.22643	8.6	42556.147	1.0			34714.22643	1.8	42556.147	0.1
	5	42556.147	1.0	51886.965	0.1			42556.147	0.1	51886.965	0.0
48	1	31826.952	26.3	31826.952	33.0			31826.952	38.1	31826.952	46.2
	2	43692.384	24.2	0.0	31.4			43692.384	24.4	0.0	33.0
	3	30656.087	19.5	30113.99	24.1			30656.087	19.0	30113.99	18.4
	4	30113.99	17.0	30656.087	3.9			30113.99	14.3	53310.101	0.2
	5	51483.98	2.5	53310.101	0.6			51483.98	0.8	58390.9	0.2
49	1	2212.599	37.1	2212.599	39.2			2212.599	45.8	2212.599	47.4
	2	0.0	21.0	0.0	36.1			0.0	20.6	0.0	33.4
	3	24372.957	12.5	31816.982	1.8			24372.957	10.6	31816.982	1.5
	4	31816.982	1.8	32115.251	1.7			31816.982	1.5	32115.251	1.4
	5	32115.251	1.7	34977.678	1.0			32115.251	1.4	34977.678	0.8
50	1	17162.499	34.2	1691.806	29.5			17162.499	53.6	1691.806	29.2
	2	34640.758	11.7	0.0	22.6			34640.758	8.9	0.0	22.1
	3	34914.282	10.7	17162.499	8.0			34914.282	8.0	17162.499	14.3
	4	8612.955	3.8	3427.673	7.0			8612.955	2.6	3427.673	6.1
	5	38628.876	3.4	8612.955	3.5			38628.876	2.4	8612.955	3.2
51	1	43249.337	29.4	0.0	57.5			43249.337	36.7	0.0	63.8
	2	45945.34	12.2	8512.125	6.2			45945.34	11.9	8512.125	5.8
	3	46991.058	8.8	9854.018	3.8			46991.058	8.2	9854.018	3.3
	4	48332.424	5.8	18464.202	3.1			48332.424	5.3	18464.202	2.8
	5	18464.202	5.1	16395.359	2.2			18464.202	4.7	16395.359	1.7
52	1	44253.0	39.9	0.0	62.7			44253.0	57.7	0.0	78.9
	2	46652.738	18.2	54160.094	2.3			46652.738	16.1	54160.094	1.4
	3	23198.392	4.7	54199.122	2.3			23198.392	3.3	54199.122	1.4
	4	54160.094	1.8	54535.345	2.1			54160.094	1.2	54535.345	1.2
	5	54199.122	1.8	55677.722	1.6			54199.122	1.2	55677.722	0.9
53	1	54633.46	35.3	0.0	60.3			54633.46	58.1	0.0	82.5
	2	56092.881	22.5	7602.97	4.7			56092.881	22.7	7602.97	2.2
	3	60896.243	4.7	63186.758	2.4			60896.243	2.4	63186.758	1.1
	4	61819.779	3.5	60896.243	1.6			61819.779	1.7	60896.243	0.8
	5	63186.758	2.3	64906.29	1.5			63186.758	1.1	64906.29	0.7

(continued)

Table B.1 (continued)

$A^+ + K \rightarrow A(k) + K^+$

| (Z) | # | \multicolumn{4}{c}{40 keV} | | | | \multicolumn{4}{c}{5 keV} | | | |

(Z)	#	40 keV Initial (0.0 cm) Level (cm⁻¹)	40 keV Initial %	40 keV Final (120.0 cm) Level (cm⁻¹)	40 keV Final %	5 keV Initial (0.0 cm) Level (cm⁻¹)	5 keV Initial %	5 keV Final (120.0 cm) Level (cm⁻¹)	5 keV Final %
54	1	67067.547	38.8	67067.547	41.7	67067.547	62.8	67067.547	63.9
	2	68045.156	29.5	0.0	33.4	68045.156	28.6	0.0	29.6
	3	76196.767	2.2	76196.767	1.6	76196.767	0.6	76196.767	0.5
	4	77185.041	1.7	78119.798	1.3	77185.041	0.5	78119.798	0.4
	5	77269.145	1.6	78403.061	1.3	77269.145	0.5	78403.061	0.4
55	1	0.0	80.2	0.0	88.0	0.0	97.3	0.0	98.4
	2	11178.26816	2.6	18535.5286	0.5	11178.26816	0.4	18535.5286	0.1
	3	11732.3071	2.2	22588.821	0.2	11732.3071	0.3	22588.821	0.0
	4	14499.2568	1.1	22631.6863	0.2	14499.2568	0.2	22631.6863	0.0
	5	14596.84232	1.1	24317.1494	0.2	14596.84232	0.2	24317.1494	0.0
56	1	9033.966	18.0	9033.966	21.4	9033.966	37.0	9033.966	38.4
	2	9215.501	17.5	9215.501	20.8	9215.501	31.2	9215.501	32.3
	3	9596.533	16.3	9596.533	19.1	9596.533	21.6	9596.533	22.4
	4	11395.35	10.7	12266.024	9.7	11395.35	4.1	12266.024	2.2
	5	12266.024	8.3	12636.623	8.8	12266.024	2.1	12636.623	1.6
60	1	9692.277	1.1	9692.277	1.3	9692.277	3.9	9692.277	4.6
	2	9814.683	1.1	9814.683	1.3	9814.683	3.8	9814.683	4.5
	3	9927.387	1.1	9939.704	1.3	9927.387	3.7	9939.704	4.3
	4	9939.704	1.1	9115.092	1.2	9939.704	3.7	9115.092	4.2
	5	10004.583	1.1	10376.842	1.2	10004.583	3.6	8800.392	3.6
62	1	10801.1	2.2	10801.1	2.8	10801.1	15.9	10801.1	17.6
	2	11044.9	2.2	11044.9	2.8	11044.9	14.6	11044.9	16.2
	3	11406.5	2.2	11406.5	2.8	11406.5	11.9	11406.5	13.2
	4	11877.5	2.1	11877.5	2.7	11877.5	8.2	11877.5	9.1
	5	12313.11	2.0	12445.35	2.5	12313.11	5.5	12445.35	5.3
63	1	12923.72	7.3	0.0	23.5	12923.72	20.2	0.0	18.7
	2	13048.9	7.1	15137.72	8.1	13048.9	17.8	12923.72	10.1
	3	13222.04	6.9	15248.76	7.9	13222.04	14.9	15137.72	9.2
	4	13457.21	6.5	15421.25	7.5	13457.21	11.7	13048.9	8.9
	5	13778.68	6.1	15680.28	6.9	13778.68	8.4	15248.76	8.4
64	1	14777.975	1.2	15744.826	1.8	14669.148	4.1	14777.975	5.2
	2	14669.148	1.2	15758.277	1.8	14777.975	4.1	14253.948	5.0
	3	15121.22	1.2	15758.651	1.8	14298.311	4.0	15121.22	4.7
	4	15173.639	1.2	15833.797	1.8	14253.948	3.9	15289.035	4.3
	5	15174.0	1.2	15989.463	1.7	15121.22	3.7	15744.826	4.0
66	1	12892.76	3.3	0.0	6.6	12892.76	11.5	13495.932	13.2
	2	12655.13	3.3	13495.932	4.5	12655.13	11.2	12298.551	11.3
	3	13495.932	3.2	12298.551	3.9	13495.932	9.8	12007.121	9.4
	4	12298.551	3.2	13952.001	3.8	12298.551	9.7	12892.76	9.3
	5	13952.001	3.1	12007.121	3.8	12007.121	8.1	12655.13	9.0
67	1	13094.42	11.3	10695.75	9.3	13094.42	51.9	13094.42	47.2
	2	15081.12	10.4	13094.42	7.4	15081.12	23.6	15081.12	21.4
	3	15792.13	9.2	15081.12	6.8	15792.13	11.4	10695.75	10.4
	4	10695.75	7.6	15792.13	6.1	10695.75	6.1	15792.13	10.3
	5	8605.16	4.1	8605.16	5.1	8605.16	1.0	8605.16	1.7

(continued)

Table B.1 (continued)

$A^+ + K \rightarrow A(k) + K^+$

(Z)	#	40 keV				5 keV			
		Initial (0.0 cm)		Final (120.0 cm)		Initial (0.0 cm)		Final (120.0 cm)	
		Level (cm^{-1})	%	Level (cm^{-1})	%	Level (cm^{-1})	%	Level (cm^{-1})	%
68	1	13097.906	8.3	13097.906	5.5	13097.906	53.5	13097.906	47.9
	2	12377.534	7.4	10750.982	5.2	12377.534	26.8	12377.534	23.9
	3	10750.982	5.0	12377.534	4.9	10750.982	5.2	10750.982	7.4
	4	19362.105	3.1	0.0	3.1	19362.105	1.2	19362.105	1.4
	5	20166.13	2.3	19362.105	2.6	20166.13	0.7	20166.13	0.8
69	1	20406.84	11.8	20406.84	8.4	20406.84	18.8	20406.84	13.8
	2	8771.243	9.7	8771.243	8.1	8771.243	15.1	8771.243	13.0
	3	21799.38	7.0	22419.764	7.9	21799.38	7.5	22419.764	7.8
	4	22419.764	5.5	22559.502	7.5	22419.764	5.3	22559.502	7.2
	5	22559.502	5.3	22742.777	7.0	22559.502	4.9	22742.777	6.6
70	1	17288.439	35.8	17288.439	36.9	17288.439	64.1	17288.439	64.2
	2	17992.007	30.8	0.0	32.9	17992.007	29.6	0.0	29.8
	3	19710.388	18.6	19710.388	19.2	19710.388	5.2	19710.388	5.2
	4	24489.102	3.1	24489.102	3.2	24489.102	0.2	24489.102	0.2
	5	24751.948	2.8	24751.948	2.9	24751.948	0.2	24751.948	0.2
71	1	7476.39	41.3	7476.39	41.4	7476.39	93.6	7476.39	93.6
	2	4136.13	17.4	4136.13	18.0	4136.13	3.6	4136.13	3.6
	3	1993.92	8.4	0.0	13.6	1993.92	0.8	1993.92	1.0
	4	0.0	4.3	1993.92	11.5	0.0	0.3	0.0	0.9
	5	17427.28	3.8	18851.31	2.2	17427.28	0.2	18851.31	0.1
72	1	19791.3	4.0	21738.7	4.9	19791.3	14.1	19292.69	12.3
	2	20784.87	4.0	17901.28	4.3	20784.87	11.2	20960.09	10.9
	3	20908.42	3.9	22450.56	4.2	19292.69	11.0	20784.87	10.0
	4	20960.09	3.9	20960.09	4.1	20908.42	10.2	19791.3	9.4
	5	19292.69	3.9	19292.69	4.0	20960.09	9.8	20908.42	9.1
73	1	26022.74	1.8	2010.1	4.9	25926.34	4.3	25181.12	6.9
	2	25926.34	1.8	12234.76	4.1	26022.74	4.3	2010.1	6.8
	3	26219.62	1.8	0.0	3.9	26219.62	4.1	25926.34	5.9
	4	26363.69	1.8	11243.63	3.6	25512.63	4.0	26022.74	5.8
	5	25512.63	1.8	25181.12	3.5	25478.3	4.0	0.0	4.9
74	1	28392.7	1.8	2951.29	20.3	28392.7	4.4	2951.29	25.8
	2	28720.88	1.8	6219.33	7.2	28347.6	4.4	1670.29	8.6
	3	28347.6	1.8	1670.29	6.1	28291.88	4.3	6219.33	5.7
	4	28291.88	1.8	0.0	4.3	28233.44	4.3	28347.6	5.3
	5	28797.24	1.8	3325.53	3.5	28204.2	4.3	28233.44	5.2
77	1	37446.13	3.3	0.0	28.4	37446.13	9.6	0.0	36.4
	2	37515.31	3.3	4078.94	9.0	37515.31	9.5	4078.94	13.8
	3	37692.75	3.3	2834.98	6.7	37692.75	9.0	37446.13	11.5
	4	37871.69	3.3	5784.62	6.0	37871.69	8.2	11831.09	10.7
	5	38120.94	3.2	11831.09	4.3	38120.94	6.9	5784.62	7.8

(continued)

Table B.1 (continued)

$A^+ + K \rightarrow A(k) + K^+$

A^+		40 keV				5 keV			
		Initial (0.0 cm)		Final (120.0 cm)		Initial (0.0 cm)		Final (120.0 cm)	
(Z)	#	Level (cm^{-1})	%	Level (cm^{-1})	%	Level (cm^{-1})	%	Level (cm^{-1})	%
79	1	41174.613	26.8	0.0	51.0	41174.613	47.2	0.0	82.3
	2	37358.991	24.2	42163.53	21.1	37358.991	35.1	42163.53	15.1
	3	42163.53	21.1	45537.195	6.4	42163.53	15.1	45537.195	0.8
	4	45537.195	6.4	46174.979	5.0	45537.195	0.8	46174.979	0.5
	5	46174.979	5.0	46379.0	4.6	46174.979	0.5	46379.0	0.5
80	1	54068.6829	39.5	44042.909	51.1	54068.6829	46.1	44042.909	64.4
	2	44042.909	35.8	0.0	30.7	44042.909	43.8	0.0	28.2
	3	39412.237	7.5	37644.982	11.2	39412.237	3.6	37644.982	4.6
	4	37644.982	4.4	71336.005	0.3	37644.982	1.8	71336.005	0.1
	5	62350.325	2.1	70932.2	0.2	62350.325	0.7	70932.2	0.1
81	1	7792.7	73.7	7792.7	75.2	7792.7	86.1	7792.7	86.6
	2	26477.5	9.1	0.0	17.1	26477.5	4.7	0.0	9.1
	3	0.0	7.1	34159.9	1.1	0.0	3.7	34159.9	0.6
	4	34159.9	1.1	35161.1	0.9	34159.9	0.6	35161.1	0.5
	5	35161.1	0.9	41368.1	0.3	35161.1	0.5	41368.1	0.1
82	1	45443.171	8.5	10650.3271	15.2	45443.171	7.8	10650.3271	13.9
	2	46060.8364	7.4	7819.2626	7.5	46060.8364	6.9	7819.2626	7.3
	3	46068.4385	7.4	0.0	7.4	46068.4385	6.8	0.0	6.8
	4	46328.6668	7.0	52101.66	2.2	46328.6668	6.5	52101.66	2.2
	5	52101.66	2.2	52311.315	2.1	52101.66	2.2	52311.315	2.1
83	1	21660.914	73.5	21660.914	76.6	21660.914	98.0	21660.914	98.3
	2	15437.501	9.0	15437.501	9.4	15437.501	0.8	15437.501	0.8
	3	32588.221	6.6	33164.805	5.6	32588.221	0.5	33164.805	0.4
	4	33164.805	5.4	0.0	3.1	33164.805	0.4	0.0	0.2
	5	11419.039	2.6	11419.039	3.0	11419.039	0.2	11419.039	0.2
87	1	0.0	94.7	0.0	97.6	0.0	99.7	0.0	99.9
	2	12237.409	1.0	16229.87	0.4	12237.409	0.1	16229.87	0.0
	3	13923.998	0.6	16429.64	0.4	13923.998	0.0	16429.64	0.0
	4	16229.87	0.4	19739.98	0.2	16229.87	0.0	19739.98	0.0
	5	16429.64	0.3	30936.325	0.0	16429.64	0.0	30936.325	0.0

B.2 Table: Relative Atomic Populations Following Neutralization with Na

See Table B.2.

Table B.2 Atomic population distribution simulation results for neutralisation of an ion beam of elements A = 1–90 by free Na atoms. Atomic data sourced from the NIST atomic database [1]. Only the 5 most populous states are listed in this table. ⋆—these light elements are outside of the intermediate velocity region at 40 keV. Some elements were not included if insufficient level or transition data was available to provide reliable values for these elements. The columns for 0 cm and 120 cm flight distance give respectively the initial populations after charge exchange at 0 cm and the final populations after a further 120 cm of atom flight at the corresponding beam energy

$A^+ + Na \rightarrow A(k) + Na^+$

| A^+ | | 40 keV | | | | 5 keV | | | |
| | | Initial (0.0 cm) | | Final (120.0 cm) | | Initial (0.0 cm) | | Final (120.0 cm) | |
(Z)	#	Level (cm^{-1})	%	Level (cm^{-1})	%	Level (cm^{-1})	%	Level (cm^{-1})	%
1⋆	1	107965.04916	1.0	0.0	41.5	82258.91911	4.4	0.0	60.8
	2	107965.04971	1.0	82258.9544	1.8	82258.9544	4.4	82258.9544	4.9
	3	107965.05487	1.0	107440.43933	1.0	82259.158	4.4	107440.4385	0.8
	4	107965.05488	1.0	108324.72418	1.0	82259.285	4.4	107440.43933	0.8
	5	107965.05677	1.0	108324.72416	1.0	97492.2112	1.8	107965.04916	0.7
2⋆	1	166277.44014	1.2	159855.97433	52.9	159855.97433	16.4	159855.97433	75.8
	2	169086.76647	1.2	0.0	9.1	166277.44014	10.4	0.0	7.0
	3	169086.8429	1.2	193921.61495	0.7	169086.76647	7.5	166277.44014	4.3
	4	169087.83081	1.2	193921.61772	0.7	169086.8429	7.5	193921.61495	0.3
	5	159855.97433	1.2	193921.62024	0.7	169087.83081	7.5	193921.61772	0.3
3⋆	1	0.0	4.4	0.0	52.5	0.0	53.2	0.0	90.2
	2	14903.66	4.1	42003.3	0.7	14903.66	12.5	42003.3	0.2
	3	14904.0	4.1	42298.0	0.7	14904.0	12.5	42298.0	0.1
	4	27206.12	2.4	42389.0	0.7	27206.12	1.4	42389.0	0.1
	5	30925.38	1.8	42389.0	0.7	30925.38	0.7	42389.0	0.1
4	1	42565.35	4.2	21978.28	64.2	42565.35	26.8	21978.28	44.8
	2	21981.27	3.1	21978.925	15.5	21981.27	15.0	21978.925	21.6
	3	21978.925	3.1	0.0	6.3	21978.925	15.0	21981.27	18.2
	4	21978.28	3.1	21981.27	5.9	21978.28	15.0	0.0	13.0
	5	52080.94	2.9	56882.43	1.6	52080.94	3.3	56882.43	0.8
5	1	28647.43	7.0	0.0	37.1	28647.43	30.6	28647.43	30.6
	2	28652.07	7.0	15.287	14.4	28652.07	30.6	28652.07	30.5
	3	28658.4	7.0	28647.43	6.9	28658.4	30.5	28658.4	30.5
	4	40039.6907	5.2	28652.07	6.9	40039.6907	2.6	0.0	5.7
	5	47856.809	3.2	28658.4	6.9	15.287	0.5	15.287	1.3
6	1	60333.43	2.2	16.4	36.7	60333.43	11.7	16.4	48.3
	2	60352.63	2.1	0.0	21.1	60352.63	11.7	0.0	29.2
	3	60393.14	2.1	43.4	13.4	60393.14	11.6	43.4	9.2
	4	61981.82	2.0	10192.63	6.0	61981.82	7.9	10192.63	5.6
	5	64086.92	1.8	33735.2	0.8	33735.2	4.9	33735.2	2.6

(continued)

Table B.2 (continued)

$A^+ + Na \rightarrow A(k) + Na^+$

A$^+$		40 keV				5 keV			
		Initial (0.0 cm)		Final (120.0 cm)		Initial (0.0 cm)		Final (120.0 cm)	
(Z)	#	Level (cm^{-1})	%	Level (cm^{-1})	%	Level (cm^{-1})	%	Level (cm^{-1})	%
7	1	83284.07	2.4	0.0	54.7	83284.07	16.2	0.0	82.8
	2	83317.83	2.4	19224.464	8.0	83317.83	16.0	19224.464	6.1
	3	83364.62	2.4	19233.177	5.1	83364.62	15.8	19233.177	5.6
	4	86137.35	2.2	28838.92	1.1	86137.35	7.5	28838.92	0.2
	5	86220.51	2.2	106868.635	1.0	86220.51	7.3	106868.635	0.2
8	1	73768.2	6.1	73768.2	46.8	73768.2	54.6	73768.2	72.3
	2	76794.978	5.5	0.0	19.9	76794.978	23.1	0.0	21.8
	3	86625.757	2.7	158.265	4.1	86625.757	1.5	158.265	1.1
	4	86627.778	2.7	102865.655	0.9	86627.778	1.5	102865.655	0.1
	5	86631.454	2.7	226.977	0.7	86631.454	1.5	105441.645	0.1
9	1	102405.71	6.0	102405.71	19.1	102405.71	23.4	102405.71	29.8
	2	102680.44	6.0	102680.44	16.0	102680.44	22.0	102680.44	27.0
	3	102840.38	6.0	102840.38	10.8	102840.38	21.2	102840.38	24.9
	4	104731.05	5.7	104731.05	7.8	104731.05	12.8	104731.05	8.2
	5	105056.28	5.6	105056.28	5.1	105056.28	11.7	105056.28	6.9
10	1	134041.84	6.6	0.0	25.2	134041.84	26.4	0.0	44.9
	2	134459.2871	6.6	134041.84	22.1	134459.2871	24.9	134041.84	33.4
	3	134818.6405	6.6	134818.6405	7.1	134818.6405	23.5	134818.6405	18.6
	4	135888.7173	6.5	158601.1152	2.0	135888.7173	18.9	158601.1152	0.1
	5	148257.7898	2.7	159379.9935	0.8	148257.7898	0.5	159379.9935	0.1
11	1	0.0	26.4	0.0	72.9	0.0	95.5	0.0	98.8
	2	16956.17025	8.4	37059.54	0.5	16956.17025	0.8	37059.54	0.0
	3	16973.36619	8.4	38400.9	0.4	16973.36619	0.8	38400.9	0.0
	4	25739.999	2.5	38400.904	0.4	25739.999	0.1	38400.904	0.0
	5	29172.837	1.5	38401.147	0.4	29172.837	0.1	38401.147	0.0
12	1	21850.405	13.4	21870.464	27.8	21850.405	32.6	21850.405	33.3
	2	21870.464	13.4	21850.405	26.0	21870.464	32.5	21870.464	33.0
	3	21911.178	13.4	21911.178	22.0	21911.178	32.3	21911.178	32.7
	4	35051.264	5.3	0.0	6.3	35051.264	0.5	0.0	0.4
	5	0.0	2.9	46403.065	2.3	0.0	0.2	46403.065	0.1
13	1	112.061	18.4	0.0	41.5	112.061	46.9	0.0	48.6
	2	0.0	18.3	112.061	30.3	0.0	45.1	112.061	48.0
	3	25347.756	5.8	29020.41	3.5	25347.756	1.2	29020.41	0.6
	4	29020.41	3.5	29066.96	3.5	29020.41	0.6	29066.96	0.5
	5	29066.96	3.5	29142.78	3.4	29066.96	0.5	29142.78	0.5
14	1	33326.053	8.5	77.115	27.9	15394.37	32.8	33326.053	25.0
	2	15394.37	7.6	0.0	22.6	33326.053	27.6	77.115	24.4
	3	39683.163	4.2	223.157	8.3	39683.163	3.5	0.0	18.3
	4	39760.285	4.2	6298.85	5.3	39760.285	3.4	15394.37	7.6
	5	39955.053	4.0	33326.053	5.2	39955.053	3.2	223.157	5.7

(continued)

Table B.2 (continued)

$A^+ + Na \rightarrow A(k) + Na^+$

		40 keV				5 keV			
A^+		Initial (0.0 cm)		Final (120.0 cm)		Initial (0.0 cm)		Final (120.0 cm)	
(Z)	#	Level (cm^{-1})	%	Level (cm^{-1})	%	Level (cm^{-1})	%	Level (cm^{-1})	%
15	1	55939.421	4.2	0.0	30.0	55939.421	9.9	0.0	50.6
	2	56090.626	4.1	11361.02	6.5	56090.626	9.5	11361.02	8.4
	3	56339.656	4.0	65788.455	1.9	56339.656	8.7	65788.455	1.4
	4	57876.574	3.3	65585.13	1.9	57876.574	5.5	65585.13	1.4
	5	58174.366	3.2	65450.125	1.9	58174.366	5.1	65450.125	1.4
16	1	52623.64	10.8	0.0	38.5	52623.64	36.5	0.0	47.6
	2	55330.811	7.8	52623.64	32.1	55330.811	14.9	52623.64	31.7
	3	22179.954	3.5	63446.065	13.6	22179.954	4.3	63446.065	8.0
	4	63446.065	2.6	396.055	6.9	63446.065	1.9	396.055	8.0
	5	63457.142	2.6	573.64	2.6	63457.142	1.9	573.64	1.3
17	1	71958.363	7.7	0.0	42.4	71958.363	22.5	0.0	67.2
	2	72488.568	7.3	71958.363	12.2	72488.568	18.3	882.3515	17.1
	3	72827.038	7.0	882.3515	12.1	72827.038	16.1	71958.363	6.1
	4	74225.846	5.9	87979.49	1.1	74225.846	9.7	87979.49	0.4
	5	74865.667	5.4	88080.042	1.1	74865.667	7.8	88080.042	0.4
18	1	93143.76	11.3	0.0	29.3	93143.76	30.0	93143.76	40.8
	2	93750.5978	10.5	93143.76	28.6	93750.5978	22.9	0.0	36.6
	3	94553.6652	9.5	94553.6652	9.6	94553.6652	16.1	94553.6652	12.8
	4	95399.8276	8.5	112750.153	1.1	95399.8276	11.4	112750.153	0.3
	5	104102.099	2.1	113716.555	0.9	104102.099	0.8	113716.555	0.3
19	1	0.0	36.5	0.0	65.5	0.0	87.8	0.0	94.0
	2	12985.18572	5.4	30617.31	0.4	12985.18572	1.2	30617.31	0.1
	3	13042.89603	5.4	30617.31	0.4	13042.89603	1.1	30617.31	0.1
	4	21026.551	1.5	33910.42	0.3	21026.551	0.3	33910.42	0.0
	5	21534.68	1.4	33910.42	0.3	21534.68	0.3	33910.42	0.0
20	1	15157.901	10.3	15210.063	14.9	15157.901	21.2	15157.901	25.8
	2	15210.063	10.2	15157.901	14.4	15210.063	20.7	15210.063	24.8
	3	15315.943	10.1	15315.943	14.0	15315.943	19.7	15315.943	23.9
	4	0.0	8.7	20335.36	9.3	0.0	18.7	0.0	8.5
	5	20335.36	5.1	20349.26	8.3	20335.36	2.7	20335.36	4.0
21	1	11677.38	3.3	0.0	12.3	11519.99	11.8	11519.99	13.9
	2	11610.28	3.3	168.34	10.6	11557.69	11.8	11557.69	13.9
	3	11557.69	3.3	11519.99	6.2	11610.28	11.8	11610.28	13.9
	4	11519.99	3.3	11557.69	5.7	11677.38	11.8	11677.38	13.9
	5	14926.07	3.0	11610.28	5.6	14926.07	4.6	0.0	8.5
22	1	14105.634	1.4	0.0	6.0	13981.773	5.3	15877.081	5.4
	2	14028.436	1.4	386.874	5.2	14028.436	5.3	16106.076	4.9
	3	13981.773	1.4	170.1328	4.9	14105.634	5.2	15975.631	4.7
	4	15108.111	1.4	6598.765	4.3	15108.111	4.3	15108.111	4.2
	5	15156.802	1.4	6742.756	3.8	15156.802	4.3	15156.802	4.2

(continued)

Table B.2 (continued)

$A^+ + Na \rightarrow A(k) + Na^+$

A^+		40 keV				5 keV			
		Initial (0.0 cm)		Final (120.0 cm)		Initial (0.0 cm)		Final (120.0 cm)	
(Z)	#	Level (cm^{-1})	%	Level (cm^{-1})	%	Level (cm^{-1})	%	Level (cm^{-1})	%
23	1	13801.54	1.3	2153.21	6.2	13801.54	4.7	14909.97	4.1
	2	13810.94	1.3	2220.11	5.6	13810.94	4.7	14949.37	4.0
	3	14514.76	1.3	2311.36	4.9	14514.76	3.9	15000.94	3.9
	4	14548.81	1.3	2112.28	4.3	14548.81	3.9	15062.96	3.9
	5	14909.97	1.3	2424.78	4.2	11100.59	3.4	2220.11	3.4
24	1	8307.5753	2.8	0.0	11.5	8307.5753	11.1	8307.5753	11.4
	2	8095.1842	2.8	7593.1484	6.6	8095.1842	10.0	8095.1842	10.4
	3	7927.441	2.7	7810.7795	4.2	7927.441	9.2	7927.441	9.6
	4	7810.7795	2.7	7927.441	4.2	7810.7795	8.7	7593.1484	9.4
	5	7750.7465	2.7	8095.1842	4.1	7750.7465	8.5	7810.7795	9.1
25	1	18705.37	4.3	0.0	13.6	18531.64	11.9	18705.37	14.6
	2	18531.64	4.3	18402.46	6.8	18705.37	11.8	18402.46	13.8
	3	18402.46	4.3	18705.37	6.6	18402.46	11.8	18531.64	13.5
	4	17637.15	4.2	18531.64	6.4	17637.15	10.7	17637.15	9.9
	5	17568.48	4.2	17451.52	5.6	17568.48	10.5	17568.48	9.8
26	1	22650.416	1.3	0.0	8.0	22249.429	3.2	22650.416	5.4
	2	22838.323	1.3	415.933	6.2	22650.416	3.2	23270.384	5.0
	3	22845.869	1.3	704.007	5.8	21999.13	3.2	0.0	4.4
	4	22946.816	1.3	888.132	4.7	22838.323	3.1	23244.838	4.3
	5	22996.674	1.3	19350.891	2.4	22845.869	3.1	23192.5	4.1
27	1	22475.36	2.0	0.0	7.7	21920.09	6.4	23611.78	9.1
	2	23152.57	2.0	816.0	5.9	22475.36	6.4	23855.62	8.3
	3	21920.09	2.0	1406.84	4.7	21780.47	6.4	24326.11	6.8
	4	23184.23	2.0	1809.33	3.6	21215.9	5.8	24733.28	5.6
	5	23207.76	2.0	23611.78	3.6	23152.57	5.7	25041.16	4.7
28	1	22102.325	4.5	204.787	16.0	22102.325	32.6	22102.325	16.5
	2	16017.306	3.5	0.0	13.7	16017.306	10.3	0.0	11.5
	3	15734.001	3.4	879.816	12.1	15734.001	8.8	16017.306	9.3
	4	15609.844	3.4	1713.087	6.6	15609.844	8.2	25753.553	9.0
	5	25753.553	3.3	1332.164	5.0	14728.84	5.1	204.787	8.7
29	1	13245.443	18.8	0.0	39.9	13245.443	41.7	13245.443	41.8
	2	30535.324	14.0	13245.443	19.2	11202.618	17.8	0.0	31.7
	3	11202.618	13.5	11202.618	16.7	30535.324	14.3	11202.618	18.9
	4	30783.697	13.4	39018.69	2.7	30783.697	12.9	39018.69	1.0
	5	39018.69	2.7	40909.16	1.9	39018.69	1.0	40909.16	0.7
30	1	32890.352	27.1	32890.352	29.4	32890.352	37.8	32890.352	37.9
	2	32501.421	26.4	32501.421	28.7	32501.421	32.2	32501.421	32.3
	3	32311.35	26.1	32311.35	28.4	32311.35	29.5	32311.35	29.6
	4	46745.413	5.5	46745.413	2.0	46745.413	0.2	46745.413	0.1
	5	53672.28	1.4	53672.28	1.5	53672.28	0.0	53672.28	0.0

(continued)

Table B.2 (continued)

$A^+ + Na \rightarrow A(k) + Na^+$

A^+ (Z)	#	40 keV Initial (0.0 cm) Level (cm^{-1})	%	Final (120.0 cm) Level (cm^{-1})	%	5 keV Initial (0.0 cm) Level (cm^{-1})	%	Final (120.0 cm) Level (cm^{-1})	%
31	1	826.19	39.7	826.19	41.6	826.19	58.0	826.19	58.2
	2	0.0	34.7	0.0	40.2	0.0	37.8	0.0	38.8
	3	24788.53	3.9	33044.05	0.8	24788.53	0.7	33044.05	0.1
	4	33044.05	0.8	33155.07	0.8	33044.05	0.1	33155.07	0.1
	5	33155.07	0.8	37975.768	0.4	33155.07	0.1	37975.768	0.1
32	1	16367.3332	39.9	557.1341	28.3	16367.3332	83.5	16367.3332	42.6
	2	7125.2989	7.3	0.0	18.6	7125.2989	2.4	557.1341	17.8
	3	37451.6893	6.0	16367.3332	8.8	37451.6893	1.7	0.0	12.0
	4	37702.3054	5.7	1409.9609	7.5	37702.3054	1.6	1409.9609	4.6
	5	39117.9021	4.1	7125.2989	4.9	39117.9021	1.1	7125.2989	3.7
33	1	50693.8	12.2	50693.8	16.1	50693.8	14.1	50693.8	18.4
	2	51610.2	9.8	51610.2	13.0	51610.2	10.7	51610.2	13.9
	3	52897.9	7.3	52897.9	9.6	52897.9	7.4	52897.9	9.7
	4	53135.6	6.9	55366.4	5.5	53135.6	7.0	55366.4	5.2
	5	54605.3	4.9	53135.6	4.6	18647.5	5.0	53135.6	4.6
35	1	63436.45	19.7	63436.45	23.8	63436.45	30.6	63436.45	35.4
	2	64907.19	13.9	64907.19	16.8	64907.19	17.0	64907.19	19.7
	3	66883.87	8.6	66883.87	10.4	66883.87	8.5	66883.87	9.8
	4	67183.58	8.0	67183.58	4.8	67183.58	7.7	67183.58	4.4
	5	68970.21	5.2	68970.21	3.1	68970.21	4.4	68970.21	2.6
36	1	79971.7417	23.5	79971.7417	30.2	79971.7417	36.6	79971.7417	42.6
	2	80916.768	18.8	80916.768	22.6	80916.768	23.7	80916.768	26.3
	3	85191.6166	6.6	85191.6166	5.5	85191.6166	5.1	85191.6166	4.1
	4	85846.7046	5.6	85846.7046	5.1	85846.7046	4.3	85846.7046	3.6
	5	91168.515	1.6	91168.515	1.5	91168.515	1.2	91168.515	1.0
37	1	0.0	58.5	0.0	63.3	0.0	79.6	0.0	82.3
	2	12578.95	3.0	12816.545	3.2	12578.95	1.6	12816.545	1.6
	3	12816.545	2.9	19355.203	0.9	12816.545	1.5	19355.203	0.5
	4	19355.203	0.9	19355.649	0.9	19355.203	0.5	19355.649	0.5
	5	19355.649	0.9	20132.51	0.8	19355.649	0.5	20132.51	0.4
38	1	0.0	33.9	0.0	20.1	0.0	82.4	0.0	66.7
	2	14317.507	11.6	14317.507	17.3	14317.507	3.9	14317.507	9.3
	3	14504.334	11.1	14898.545	14.4	14504.334	3.6	14898.545	7.1
	4	14898.545	10.1	14504.334	13.5	14898.545	3.1	14504.334	4.9
	5	18159.04	4.5	18159.04	7.8	18159.04	1.0	18159.04	2.9
39	1	10529.169	6.1	0.0	6.7	10529.169	20.5	10529.169	20.5
	2	10937.39	5.9	10937.39	6.2	10937.39	15.9	10937.39	16.0
	3	11078.614	5.8	10529.169	6.1	11078.614	14.5	11078.614	14.5
	4	11277.928	5.7	11078.614	5.9	11277.928	12.7	11277.928	12.7
	5	11359.757	5.6	11277.928	5.9	11359.757	12.0	11359.757	12.0

(continued)

Table B.2 (continued)

$A^+ + Na \rightarrow A(k) + Na^+$

A^+		40 keV				5 keV			
		Initial (0.0 cm)		Final (120.0 cm)		Initial (0.0 cm)		Final (120.0 cm)	
(Z)	#	Level (cm^{-1})	%	Level (cm^{-1})	%	Level (cm^{-1})	%	Level (cm^{-1})	%
42	1	16641.081	3.3	0.0	5.2	16641.081	12.8	16641.081	13.1
	2	16692.905	3.3	16692.905	3.9	16692.905	12.5	16692.905	12.9
	3	16747.72	3.3	16747.72	3.9	16747.72	12.2	16747.72	12.6
	4	16783.856	3.3	16641.081	3.9	16783.856	12.1	16783.856	12.4
	5	16784.522	3.3	16783.856	3.9	16784.522	12.1	16784.522	12.4
43	1	16287.79	2.4	16428.71	4.7	16025.15	6.1	16428.71	10.4
	2	16133.98	2.4	16874.51	4.7	16133.98	6.1	16874.51	9.1
	3	16025.15	2.4	17522.92	4.6	15770.42	6.1	17522.92	6.6
	4	16415.64	2.4	0.0	4.5	16287.79	6.0	16025.15	5.4
	5	16428.71	2.4	16287.79	2.4	15624.25	5.9	16133.98	5.4
44	1	17096.87	2.8	0.0	3.5	17096.87	12.6	17096.87	18.6
	2	17045.97	2.8	17096.87	3.2	17045.97	12.3	16240.13	12.0
	3	16712.58	2.8	16240.13	3.0	16712.58	10.6	16712.58	11.2
	4	16240.13	2.7	15550.16	2.8	16240.13	8.1	16190.61	8.3
	5	20055.71	2.7	1190.64	2.6	16190.61	7.8	15550.16	7.6
45	1	16943.5	6.9	16943.5	11.3	16943.5	26.4	16943.5	36.7
	2	16120.72	6.3	16118.69	10.3	16120.72	15.0	16118.69	20.8
	3	16118.69	6.3	0.0	9.4	16118.69	15.0	16120.72	10.4
	4	16017.94	6.2	14787.87	8.4	16017.94	13.9	16017.94	9.6
	5	14787.87	5.2	12723.07	5.7	14787.87	5.7	14787.87	7.9
46	1	25101.235	20.3	25101.235	25.2	25101.235	62.9	25101.235	63.6
	2	28213.767	18.3	28213.767	22.7	28213.767	27.0	28213.767	27.3
	3	29711.109	14.7	29711.109	18.2	29711.109	7.5	29711.109	7.5
	4	34068.977	5.5	6564.148	8.4	34068.977	0.5	6564.148	0.4
	5	35041.751	4.3	34068.977	3.7	35041.751	0.3	34068.977	0.3
47	1	29552.05741	29.6	0.0	59.1	29552.05741	34.8	0.0	62.9
	2	30242.29835	24.7	30242.29835	24.7	30242.29835	26.2	30242.29835	26.2
	3	30472.66516	23.2	34714.22643	7.9	30472.66516	23.9	34714.22643	5.7
	4	34714.22643	7.9	42556.147	1.5	34714.22643	5.7	42556.147	0.9
	5	0.0	4.3	51886.965	0.3	0.0	2.9	51886.965	0.2
48	1	31826.952	31.8	31826.952	32.7	30656.087	36.4	0.0	36.3
	2	30656.087	31.7	30113.99	32.0	31826.952	33.6	31826.952	33.7
	3	30113.99	30.8	0.0	28.5	30113.99	29.9	30113.99	29.9
	4	43692.384	2.4	30656.087	5.1	43692.384	0.0	30656.087	0.2
	5	51483.98	0.4	58390.9	0.1	51483.98	0.0	53310.101	0.0
49	1	2212.599	56.8	2212.599	57.3	2212.599	83.3	2212.599	83.4
	2	0.0	36.9	0.0	38.1	0.0	16.2	0.0	16.3
	3	24372.957	1.0	31816.982	0.2	24372.957	0.1	31816.982	0.0
	4	31816.982	0.2	32115.251	0.2	31816.982	0.0	32115.251	0.0
	5	32115.251	0.2	34977.678	0.1	32115.251	0.0	34977.678	0.0

(continued)

Table B.2 (continued)

$A^+ + Na \rightarrow A(k) + Na^+$

| A^+ | | 40 keV | | | | 5 keV | | | |
| (Z) | # | Initial (0.0 cm) | | Final (120.0 cm) | | Initial (0.0 cm) | | Final (120.0 cm) | |
		Level (cm^{-1})	%	Level (cm^{-1})	%	Level (cm^{-1})	%	Level (cm^{-1})	%
50	1	17162.499	67.5	17162.499	26.1	17162.499	99.1	17162.499	95.4
	2	8612.955	12.4	1691.806	15.7	8612.955	0.4	8612.955	1.3
	3	3427.673	3.5	8612.955	14.5	3427.673	0.1	1691.806	0.9
	4	1691.806	2.4	3427.673	13.9	1691.806	0.1	3427.673	0.8
	5	0.0	1.7	0.0	11.4	0.0	0.0	0.0	0.7
51	1	18464.202	35.9	18464.202	29.5	18464.202	46.4	18464.202	39.9
	2	16395.359	21.2	0.0	19.1	16395.359	21.6	16395.359	18.5
	3	43249.337	6.3	16395.359	17.5	43249.337	5.0	0.0	15.7
	4	9854.018	4.8	9854.018	5.5	9854.018	3.8	9854.018	4.5
	5	8512.125	3.6	8512.125	5.4	8512.125	2.9	8512.125	4.5
52	1	23198.392	50.3	0.0	30.8	23198.392	69.4	23198.392	37.2
	2	44253.0	11.0	23198.392	21.0	44253.0	7.5	0.0	26.8
	3	46652.738	5.7	54160.094	2.2	46652.738	3.6	54160.094	1.7
	4	10557.877	2.7	54199.122	2.1	10557.877	1.7	54199.122	1.7
	5	4750.712	1.1	54535.345	2.0	4750.712	0.7	54535.345	1.6
53	1	54633.46	20.8	0.0	36.2	54633.46	24.5	0.0	41.2
	2	56092.881	13.6	7602.97	5.0	56092.881	14.8	7602.97	5.0
	3	60896.243	3.8	63186.758	2.4	60896.243	3.8	63186.758	2.4
	4	61819.779	3.1	64906.29	1.7	61819.779	3.1	64906.29	1.7
	5	63186.758	2.3	64989.994	1.6	63186.758	2.3	64989.994	1.6
54	1	67067.547	24.3	67067.547	27.9	67067.547	27.5	67067.547	31.2
	2	68045.156	18.2	0.0	23.4	68045.156	19.3	0.0	24.3
	3	76196.767	2.2	76196.767	1.8	76196.767	2.2	76196.767	1.8
	4	77185.041	1.8	78403.061	1.7	77185.041	1.8	78403.061	1.6
	5	77269.145	1.8	78119.798	1.6	77269.145	1.7	78119.798	1.5
55	1	0.0	56.8	0.0	68.1	0.0	62.1	0.0	72.8
	2	11178.26816	3.3	18535.5286	0.9	11178.26816	2.9	18535.5286	0.8
	3	11732.3071	2.9	22588.821	0.5	11732.3071	2.6	22588.821	0.4
	4	14499.2568	1.7	22631.6863	0.5	14499.2568	1.6	22631.6863	0.4
	5	14596.84232	1.7	24317.1494	0.4	14596.84232	1.5	24317.1494	0.3
56	1	0.0	48.8	0.0	27.0	0.0	98.3	0.0	95.5
	2	9033.966	9.1	9033.966	15.7	9033.966	0.4	9033.966	1.2
	3	9215.501	8.6	9215.501	14.8	9215.501	0.3	9215.501	1.1
	4	9596.533	7.7	9596.533	12.8	9596.533	0.3	9596.533	0.9
	5	11395.35	4.5	12266.024	5.8	11395.35	0.1	12266.024	0.3
60	1	3681.696	3.8	3681.696	3.3	3681.696	28.7	3681.696	28.0
	2	2366.597	3.7	2366.597	3.2	2366.597	26.3	2366.597	25.6
	3	5048.602	3.4	5048.602	2.9	5048.602	11.5	5048.602	11.2
	4	1128.056	3.3	1128.056	2.8	1128.056	10.9	1128.056	10.6
	5	0.0	2.7	0.0	2.5	0.0	4.2	0.0	4.1

(continued)

Table B.2 (continued)

$A^+ + Na \rightarrow A(k) + Na^+$

| A^+ | | 40 keV | | | | 5 keV | | | |
| | | Initial (0.0 cm) | | Final (120.0 cm) | | Initial (0.0 cm) | | Final (120.0 cm) | |
(Z)	#	Level (cm^{-1})	%	Level (cm^{-1})	%	Level (cm^{-1})	%	Level (cm^{-1})	%
62	1	4020.66	7.1	4020.66	7.1	4020.66	37.9	4020.66	37.8
	2	3125.46	6.8	3125.46	6.7	3125.46	27.7	3125.46	27.6
	3	2273.09	6.1	2273.09	6.1	2273.09	14.5	2273.09	14.5
	4	1489.55	5.4	1489.55	5.4	1489.55	7.2	1489.55	7.2
	5	811.92	4.7	811.92	4.7	811.92	3.9	811.92	3.9
63	1	0.0	19.1	0.0	46.4	0.0	60.7	0.0	82.2
	2	12923.72	5.6	15137.72	4.7	12923.72	3.4	15137.72	1.7
	3	13048.9	5.4	15248.76	4.6	13048.9	3.2	15248.76	1.7
	4	13222.04	5.1	15421.25	4.5	13222.04	2.9	15421.25	1.6
	5	13457.21	4.8	15680.28	4.0	13457.21	2.7	15680.28	1.4
64	1	8498.434	2.8	10222.233	4.0	7947.294	9.3	7653.927	13.8
	2	7947.294	2.8	10359.905	3.9	8498.434	9.1	7562.457	13.3
	3	7653.927	2.8	10576.41	3.8	7653.927	8.6	7426.71	12.5
	4	7562.457	2.8	10883.505	3.6	7562.457	8.3	7234.91	11.2
	5	7480.348	2.8	7653.927	3.6	7480.348	8.0	6976.508	9.4
66	1	7050.603	8.5	7050.603	9.3	7050.603	35.2	7050.603	35.5
	2	7565.61	8.2	7565.61	9.0	7565.61	25.9	7565.61	26.2
	3	8519.21	7.3	8519.21	8.1	8519.21	11.5	8519.21	11.6
	4	4134.222	6.7	4134.222	7.8	4134.222	9.1	4134.222	9.2
	5	9211.591	6.5	9211.591	7.1	9211.591	6.0	9211.591	6.0
67	1	8605.16	22.6	8605.16	24.5	8605.16	49.2	8605.16	49.6
	2	5419.7	21.3	5419.7	23.0	5419.7	40.5	5419.7	40.8
	3	10695.75	15.2	10695.75	16.5	10695.75	7.0	10695.75	7.0
	4	13094.42	7.5	0.0	6.7	13094.42	1.1	0.0	0.7
	5	0.0	5.3	13094.42	4.4	0.0	0.6	13094.42	0.6
68	1	6958.329	18.7	6958.329	20.9	6958.329	69.7	6958.329	74.3
	2	10750.982	14.0	10750.982	15.6	5035.193	13.0	10750.982	11.6
	3	5035.193	13.9	5035.193	9.8	10750.982	10.9	5035.193	8.8
	4	12377.534	9.2	12377.534	6.5	12377.534	2.5	12377.534	1.7
	5	13097.906	7.4	13097.906	5.2	13097.906	1.4	13097.906	1.0
69	1	8771.243	63.7	8771.243	60.7	8771.243	99.1	8771.243	99.0
	2	0.0	8.9	0.0	10.5	0.0	0.3	0.0	0.3
	3	20406.84	2.5	22419.764	2.2	20406.84	0.1	22419.764	0.0
	4	21799.38	1.6	22559.502	2.1	21799.38	0.0	22559.502	0.0
	5	22419.764	1.3	20406.84	2.1	22419.764	0.0	20406.84	0.0
70	1	17288.439	26.7	0.0	36.3	17288.439	28.6	0.0	37.3
	2	0.0	23.2	17288.439	33.0	0.0	25.5	17288.439	35.8
	3	17992.007	20.9	19710.388	14.2	17992.007	20.6	19710.388	12.9
	4	19710.388	11.5	24489.102	3.2	19710.388	10.3	24489.102	2.9
	5	24489.102	2.6	24751.948	3.0	24489.102	2.3	24751.948	2.7

(continued)

Table B.2 (continued)

$A^+ + Na \rightarrow A(k) + Na^+$

		40 keV				5 keV			
A^+		Initial (0.0 cm)		Final (120.0 cm)		Initial (0.0 cm)		Final (120.0 cm)	
(Z)	#	Level (cm^{-1})	%	Level (cm^{-1})	%	Level (cm^{-1})	%	Level (cm^{-1})	%
71	1	1993.92	31.1	1993.92	31.6	1993.92	63.4	1993.92	63.4
	2	4136.13	27.5	4136.13	27.6	4136.13	21.9	4136.13	21.9
	3	0.0	24.2	0.0	25.7	0.0	13.6	0.0	13.6
	4	7476.39	11.7	7476.39	11.7	7476.39	1.0	7476.39	1.0
	5	17427.28	0.4	18851.31	0.3	17427.28	0.0	18851.31	0.0
72	1	14017.83	6.9	14092.28	9.3	14017.83	19.6	14092.28	22.9
	2	14092.28	6.9	14740.68	8.9	14092.28	19.1	14017.83	17.7
	3	14435.13	6.8	15673.33	7.8	14435.13	15.9	14740.68	15.3
	4	14541.68	6.7	14017.83	7.0	14541.68	14.8	14435.13	14.3
	5	14740.68	6.6	14435.13	6.8	14740.68	12.7	14541.68	13.3
73	1	19657.78	2.7	0.0	6.3	19657.78	10.2	0.0	12.8
	2	19178.45	2.7	20646.54	3.7	19178.45	10.1	19657.78	12.1
	3	20340.39	2.6	21153.33	3.5	20340.39	7.5	20646.54	9.9
	4	20560.26	2.6	21622.92	3.2	18504.72	7.2	19178.45	8.4
	5	18504.72	2.5	23355.41	3.2	20560.26	6.4	21153.33	6.2
74	1	22476.68	2.6	2951.29	8.9	22476.68	8.5	1670.29	19.4
	2	21453.9	2.6	1670.29	8.6	21453.9	8.4	0.0	10.2
	3	21448.76	2.6	0.0	6.4	21448.76	8.3	22476.68	9.4
	4	22773.78	2.6	4830.0	4.8	22773.78	7.3	23047.31	9.3
	5	22852.8	2.6	23047.31	4.3	22852.8	6.9	22773.78	8.1
77	1	30529.66	6.0	0.0	27.7	30529.66	32.3	30529.66	29.7
	2	32463.58	5.5	7106.61	8.5	32463.58	13.0	7106.61	16.1
	3	32513.43	5.4	5784.62	8.2	32513.43	12.4	0.0	15.5
	4	32830.78	5.1	32830.78	7.0	32830.78	8.9	5784.62	12.4
	5	33064.83	4.9	2834.98	6.5	33064.83	7.0	32830.78	11.5
79	1	37358.991	51.5	0.0	65.4	37358.991	80.6	0.0	87.3
	2	41174.613	13.8	42163.53	9.8	41174.613	6.6	42163.53	4.1
	3	42163.53	9.8	21435.191	5.5	42163.53	4.1	21435.191	2.2
	4	21435.191	5.5	45537.195	3.3	21435.191	2.2	45537.195	1.1
	5	45537.195	3.3	46174.979	2.7	45537.195	1.1	46174.979	0.9
80	1	44042.909	47.6	44042.909	48.7	44042.909	85.9	44042.909	86.0
	2	39412.237	29.9	0.0	31.4	39412.237	11.3	0.0	11.4
	3	37644.982	17.6	37644.982	18.5	37644.982	2.6	37644.982	2.6
	4	54068.6829	2.1	71336.005	0.0	54068.6829	0.1	71336.005	0.0
	5	62350.325	0.2	70932.2	0.0	62350.325	0.0	76466.936	0.0

(continued)

Table B.2 (continued)

$A^+ + Na \rightarrow A(k) + Na^+$

A^+		40 keV				5 keV			
		Initial (0.0 cm)		Final (120.0 cm)		Initial (0.0 cm)		Final (120.0 cm)	
(Z)	#	Level (cm^{-1})	%	Level (cm^{-1})	%	Level (cm^{-1})	%	Level (cm^{-1})	%
81	1	7792.7	86.3	7792.7	86.4	7792.7	99.7	7792.7	99.7
	2	0.0	12.0	0.0	12.6	0.0	0.3	0.0	0.3
	3	26477.5	0.5	34159.9	0.1	26477.5	0.0	34159.9	0.0
	4	34159.9	0.1	35161.1	0.1	34159.9	0.0	35161.1	0.0
	5	35161.1	0.1	41368.1	0.0	35161.1	0.0	46949.9	0.0
82	1	45443.171	6.2	10650.3271	11.4	45443.171	5.9	10650.3271	10.6
	2	46060.8364	5.6	7819.2626	5.7	46060.8364	5.3	7819.2626	5.6
	3	46068.4385	5.6	0.0	5.6	46068.4385	5.3	0.0	5.3
	4	46328.6668	5.3	52101.66	2.1	46328.6668	5.1	52101.66	2.1
	5	52101.66	2.1	52311.315	2.1	52101.66	2.1	52311.315	2.1
83	1	15437.501	52.6	15437.501	52.9	15437.501	90.2	15437.501	90.2
	2	21660.914	28.0	21660.914	28.1	21660.914	7.2	21660.914	7.2
	3	11419.039	16.4	11419.039	16.6	11419.039	2.5	11419.039	2.5
	4	32588.221	0.8	0.0	0.9	32588.221	0.0	0.0	0.1
	5	0.0	0.7	33164.805	0.7	0.0	0.0	33164.805	0.0
87	1	0.0	78.4	0.0	88.7	0.0	83.1	0.0	91.7
	2	12237.409	2.4	16229.87	1.4	12237.409	1.9	16229.87	1.2
	3	13923.998	1.7	16429.64	1.3	13923.998	1.3	16429.64	1.1
	4	16229.87	1.1	19739.98	0.9	16229.87	0.9	19739.98	0.9
	5	16429.64	1.1	30936.325	0.1	16429.64	0.8	30936.325	0.1

B.3   Table: Total Neutralisation Cross-Sections

See Table B.3.

Table B.3 Total cross-sections σ_{CE} for neutralisation of an ion beam of elements $A = 1-90$ by free K or Na atoms. Atomic data sourced from the NIST atomic database [1]. ‡—insufficient level data was available to provide reliable values for these elements

	Total cross-section σ_{CE} (cm^2)					Total cross-section σ_{CE} (cm^2)			
	$A^+ + K \to A + K^+$		$A^+ + Na \to A + Na^+$			$A^+ + K \to A + K^+$		$A^+ + Na \to A + Na^+$	
$A^+(Z)$	40 keV	5 keV	40 keV	5 keV	$A^+(Z)$	40 keV	5 keV	40 keV	5 keV
1	3.7e-13	1.1e-13	3.0e-13	7.0e-14	46	7.3e-14	1.5e-14	3.1e-14	1.2e-14
2	4.0e-13	5.6e-14	2.8e-13	3.1e-14	47	1.7e-14	2.5e-15	3.7e-15	8.7e-17
3	1.2e-13	1.5e-14	7.6e-14	9.8e-15	48	1.1e-14	5.9e-16	2.0e-14	2.3e-14
4	1.3e-13	1.2e-14	8.8e-14	7.5e-15	49	2.9e-15	6.2e-17	8.4e-15	1.9e-15
5	8.9e-14	2.0e-14	6.1e-14	1.6e-14	50	5.7e-15	1.5e-16	9.2e-15	7.8e-15
6	2.7e-13	3.0e-14	1.7e-13	8.4e-15	51	3.6e-15	8.0e-17	2.8e-15	7.0e-17
7	2.8e-13	4.8e-14	1.7e-13	1.2e-14	52	5.3e-15	1.6e-16	2.8e-15	8.6e-17
8	1.2e-13	1.7e-14	7.1e-14	5.3e-15	53	8.7e-15	3.6e-16	2.2e-15	4.1e-17
9	1.3e-13	3.8e-14	7.3e-14	2.0e-14	54	1.1e-14	6.7e-16	2.5e-15	4.5e-17
10	1.2e-13	2.0e-14	6.8e-14	2.3e-14	55	6.1e-15	7.3e-16	1.3e-15	2.4e-17
11	2.8e-14	2.6e-15	1.7e-14	7.1e-15	56	3.8e-14	8.6e-15	1.3e-14	8.1e-15
12	5.2e-14	9.8e-15	3.5e-14	1.9e-14	57	2.0e-14	7.4e-14	9.2e-14	3.9e-14
13	2.8e-14	9.3e-16	1.9e-14	2.8e-15	58	1.4e-12	5.3e-13	5.6e-13	2.2e-13
14	7.3e-14	8.0e-15	4.0e-14	1.7e-15	59	6.3e-13	2.8e-13	1.9e-13	3.5e-14
15	1.1e-13	5.7e-15	5.2e-14	1.0e-15	60	7.3e-13	2.7e-13	1.7e-13	2.8e-14
16	5.2e-14	4.6e-15	2.6e-14	6.3e-16	61	7.9e-14	1.0e-14	7.5e-14	3.9e-14
17	8.9e-14	1.9e-14	4.2e-14	1.7e-15	62	3.6e-13	6.5e-14	9.6e-14	2.4e-14
18	6.3e-14	2.4e-14	3.1e-14	1.6e-15	63	9.1e-14	1.2e-14	1.9e-14	6.2e-16
19	2.0e-14	9.0e-15	1.1e-14	9.1e-16	64	6.4e-13	2.6e-13	2.4e-13	9.6e-14
20	6.5e-14	2.8e-14	3.4e-14	2.5e-15	65	4.5e-13	2.6e-13	3.1e-13	1.5e-13
21	2.7e-13	1.9e-13	1.6e-13	6.5e-14	66	2.4e-13	9.5e-14	7.8e-14	2.3e-14
22	6.2e-13	2.9e-13	3.7e-13	1.4e-13	67	6.9e-14	1.9e-14	2.7e-14	8.9e-15
23	6.7e-13	3.5e-13	4.1e-13	1.5e-13	68	8.8e-14	1.1e-14	3.4e-14	9.8e-15
24	3.4e-13	9.8e-14	1.4e-13	1.2e-14	69	2.0e-14	4.8e-16	1.1e-14	8.8e-15
25	2.2e-13	1.2e-13	1.3e-13	6.7e-14	70	1.9e-14	4.6e-15	3.2e-15	6.8e-17
26	6.3e-13	2.5e-13	4.5e-13	2.5e-13	71	1.7e-14	5.7e-15	2.1e-14	1.4e-14
27	4.3e-13	1.9e-13	2.9e-13	1.2e-13	72	2.0e-13	7.5e-14	9.7e-14	4.4e-14
28	2.6e-13	1.3e-13	1.3e-13	1.6e-14	73	4.5e-13	2.5e-13	2.5e-13	9.0e-14
29	2.8e-14	5.0e-15	1.5e-14	6.2e-16	74	4.4e-13	2.5e-13	2.6e-13	9.9e-14
30	1.8e-14	1.1e-15	2.1e-14	1.6e-14	75	2.4e-13	1.4e-13	1.4e-13	5.5e-14
31	6.4e-15	1.3e-16	8.2e-15	7.6e-16	76	2.1e-13	6.0e-14	1.7e-13	5.5e-14
32	1.3e-14	3.1e-16	8.2e-15	5.3e-16	77	2.5e-13	1.1e-13	1.1e-13	2.7e-14
33	1.8e-14	6.5e-16	5.8e-15	1.1e-16	78	6.8e-14	2.8e-14	2.2e-14	8.2e-15
34	1.2e-14	1.1e-15	4.5e-15	1.1e-16	79	2.6e-14	6.2e-15	6.1e-15	2.6e-16
35	2.6e-14	3.6e-15	7.5e-15	2.1e-16	80	6.8e-15	2.7e-16	1.4e-14	5.4e-15
36	2.5e-14	6.9e-15	7.9e-15	2.1e-16	81	2.0e-15	5.4e-17	8.2e-15	9.5e-15
37	1.0e-14	6.8e-15	3.7e-15	1.3e-16	82	2.3e-16	1.9e-18	1.1e-16	3.7e-19
38	3.0e-14	4.3e-15	1.2e-14	9.6e-16	83	8.6e-15	2.0e-15	1.1e-14	2.7e-15
39	2.2e-13	1.6e-13	9.5e-14	2.2e-14	84	3.4e-15	7.0e-17	3.7e-15	1.7e-16
40	5.0e-13	2.5e-13	2.8e-13	1.1e-13	85‡				
41	5.7e-13	2.7e-13	3.6e-13	1.8e-13	86	8.7e-15	8.3e-16	9.5e-16	1.4e-17
42	4.1e-13	1.6e-13	1.9e-13	5.7e-14	87	6.6e-15	1.6e-15	7.2e-16	1.2e-17
43	4.0e-13	2.1e-13	2.6e-13	1.4e-13	88‡				
44	3.6e-13	1.4e-13	2.2e-13	5.7e-14	89	3.9e-14	1.6e-14	1.6e-14	1.1e-14
45	1.2e-13	4.1e-14	8.3e-14	1.7e-14					

Reference

1. Kramida NATA, Ralchenko Y, Reader J (2014) NIST Atomic Spectra Database. http://www.nist.gov/pml/data/asd.cfm

Curriculum Vitae

Adam R. Vernon

Contact

Nationality:	British
Place and Date of Birth:	Stafford, United Kingdom I 14 August 1993
Address:	37 Pellfield Court, Weston, Staffordshire, ST18 0JG, United Kingdom
Phone:	+44 7913 620650
email:	adam.vernon@cern.ch; adamvernon@outlook.com

Education

May 2019- Sep 2015	PhD in Nuclear Physics, **University of Manchester** **Thesis title: Evolution of proton-hole states up to $N = 82$ studied with laser spectroscopy of indium** Supervisors: Dr. Kieran T. Flanagan, Dr. Ronald F. Garcia Ruiz I Awarded
Jun 2015- Sep 2011	MEng in Nuclear Engineering, **University of Birmingham** **Thesis title: Using the $^{12}C(\alpha, \alpha)^{12}C$ reaction to investigate the $K^\pi = 0^+$ rotational band of ^{16}O associated with a $^{12}C+\alpha$ cluster structure** Supervisors: Prof. Martin Freer, Dr. Carl Wheldon *1st class degree* I 4 years highest grade in cohort
Aug 2013	'A Look at Nuclear Science and Technology' module, **coursera.org**
Aug 2012	'Inside Nuclear Energy' module, **Open University**
Aug 2011	'The story of mathematics' module, **Open University**
Jul 2011- Sep 2009	A-Levels, **Weston Road High School** Physics (A*), Mathematics (A), Further Mathematics (A), Computing (A)
Jul 2009	11 GCSEs including Maths and English

A. R. Vernon, *Collinear Resonance Ionization Spectroscopy of Neutron-Rich Indium Isotopes*, Springer Theses, https://doi.org/10.1007/978-3-030-54189-7

Selected Publications

2019   **A. R. Vernon** et al.
Simulations of atomic populations of ions $1 \leq Z \leq 89$ following charge exchange tested with collinear resonance ionization spectroscopy of indium
Spectrochim. Acta Part B doi:10.1016/j.sab.2019.02.001
A. R. Vernon et al.
Optimising the Collinear Resonance Ionization Spectroscopy experiment
Nucl. Instrum. Methods Phys. Res. B, doi:10.1016/j.nimb.2019.04.049
B. K. Sahoo, **A. R. Vernon**, R.F. Garcia Ruiz, C.L. Binnersley et al.
Analytic Response Relativistic Coupled-Cluster Theory: The first application to In isotope shifts
New Journal of Physics, doi:10.1088/1367-2630/ab66dd

2018   R.F. Garcia Ruiz, **A.R. Vernon**, C.L. Binnersley, B.K. Sahoo et al.
High-Precision Multiphoton Ionization of Accelerated Laser-Ablated Species
Phys. Rev. X. 8 (2018) 041005. doi:10.1103/PhysRevX.8.041005
Featured viewpoint article in APS physics.

2017   G. J. Farooq-Smith, **A. R. Vernon** et al.
Probing the Ga_{31} ground-state properties in the region near $Z = 28$ with high-resolution laser spectroscopy,
Phys. Rev. C, vol. 96, no. 4, p. 044324, Oct. 2017. doi:10.1103/PhysRevC.96.044324
| Other publications

Presentations, Summer Schools and Conferences

2019   Nuclear Structure and Dynamics, Venice, Italy - *invited talk*

2018   Electromagnetic Isotope Separators conference, CERN, Switzerland - *poster*
State Key Laboratory of Nuclear Physics group seminar, Peking University, China - *invited talk*
ECT* Training in Advanced Low Energy Nuclear Theory summer school, Xinxiang, China
ISOLDE workshop, CERN, Switzerland - *talk*

2017   Mazurian lakes conference on Physics, Piaski, Poland - *Poster and 10 minute talk*
STFC Nuclear Physics Summer School, Queen's University, Belfast, UK - *talk*
ISOLDE workshop, CERN, Switzerland - *talk*

2016   Euroschool on exotic beams, Mainz, Germany - *poster and 10 minute talk*
IoP nuclear physics conference, University of Liverpool, UK - *poster*
Muon spectroscopy workshop, Paul Scherrer Institut, Switzerland
ISOLDE workshop, CERN, Switzerland

2014   51st Plasma Physics Summer School, Culham Science Centre, UK

Scholarships and Awards

2020   SpringerNature outstanding PhD Thesis award

2016–2018   STFC studentship attachment at CERN
2015   EDF Nuclear Engineering prize
2013   British Non-Ferrous Metals Federation prize
2011–2015   Birmingham Academic Achievement scholarship
2011   Governor's prize for achievement in Sixth Form

Research Employment

May 2019–*Present* Research Fellow, **KU Leuven**
Research Experience

Laboratory: Ultra-high vacuum systems. Analog, digital, laser and nuclear electronics.
High-voltage. Ion sources and ion optics.
Gas systems. Ion, α, β and γ detectors.

Laser: High-power pulsed. Continuous wave. Broadband and narrowband.
Ti:Sa. Nd:YAG. Diode. Dye. Injection locking. Fiber optics.
Harmonic generation. Ablation.

Programming: Python ($>$2000 h):
Laser spectroscopy data analysis. Automated ion beam tuning.
Ion neutralisation cross section simulations.
Basic nuclear shell model and coupled cluster calculations.
Control systems and data acquisition (high-voltage, pneumatic, …)
FORTRAN, C++ & Mathematica (200–300 h):
Nuclear reactions data analysis. Undergraduate computing courses.
COMSOL Multiphysics® ($\sim$400 h)
Ion optics, thermal and vacuum simulations.
PHP, HTML, Pascal/Delphi (100–200 h)
C#, JAVA, MATLAB & MySQL (10–30 h)
LaTeX & Unix ($>$2000 h)

Mentoring: CERN summer student co-supervisor 2017-2019

Outreach: ISOLDE-CERN tours to general public/students. 2017-2019
Nucleosynthesis educational card game - Stall at INPC 2019 conference

All Publications

2020   A. R. Vernon et al.
Laser spectroscopy of indium Rydberg atom bunches by electric field ionization
Scientific Reports, Submitted
R.F. Garcia Ruiz & **A. R. Vernon**
Emergence of simple patterns in many-body systems:
from macroscopic objects to the atomic nucleus

European Physical Journal A, Accepted, arXiv:1911.04819
B. K. Sahoo, **A. R. Vernon**, R.F. Garcia Ruiz, C.L. Binnersley et al.
Analytic Response Relativistic Coupled-Cluster Theory:
The first application to In isotope shifts
New Journal of Physics (fast track communication) doi:10.1088/1367-2630/ab66dd
R.F. Garcia Ruiz, *25 co-authors* (incl. **A. R. Vernon**)
Spectroscopy of short-lived radioactive molecules:
A sensitive laboratory for new physics
Nature, doi:10.1038/s41586-020-2299-4
R. P. de Groote, *23 co-authors* (incl. **A. R. Vernon**)
Precise measurement and microscopic description of charge radii of exotic
copper isotopes:
global trends and odd-even variations
Nature Physics, doi:10.1038/s41567-020-0868-y
2019   **A. R. Vernon** et al.
Simulations of atomic populations of ions $1 \leq Z \leq 89$ following charge
exchange
tested with collinear resonance ionization spectroscopy of indium
Spectrochim. Acta Part B doi:10.1016/j.sab.2019.02.001
A. R. Vernon et al.
Optimising the Collinear Resonance Ionization Spectroscopy experiment
Nucl. Instrum. Methods Phys. Res. B, doi:10.1016/j.nimb.2019.04.049
K. Chrysalidis, *10 co-authors* (incl. **A. R. Vernon**)
First demonstration of Doppler-free 2-photon
in-source laser spectroscopy at the ISOLDE-RILIS
Nucl. Instrum. Methods Phys. Res. B, doi:10.1016/j.nimb.2019.04.020
C. M. Ricketts, *17 co-authors* (incl. **A. R. Vernon**)
A compact linear Paul trap cooler buncher for CRIS
Nucl. Instrum. Methods Phys. Res. B, doi:10.1016/j.nimb.2019.04.054
A. Koszorus, X.F. Yang, *17 co-authors* (incl. **A. R. Vernon**)
Precision measurements of the charge radii of potassium isotopes
Phys. Rev. C, doi:10.1103/PhysRevC.100.034304
A. Koszorus, *28 co-authors* (incl. **A. R. Vernon**)
Resonance ionization schemes for high resolution and high efficiency
studies of exotic nuclei at the CRIS experiment
Nucl. Instrum. Methods Phys. Res. B, doi:10.1016/j.nimb.2019.04.043
2018   R.F. Garcia Ruiz, **A.R. Vernon**, C.L. Binnersley, B.K. Sahoo et al.
High-Precision Multiphoton Ionization of Accelerated Laser-Ablated Species
Phys. Rev. X. 8 (2018) 041005. doi:10.1103/PhysRevX.8.041005
Featured viewpoint article in APS physics.
K. M. Lynch, S. G. Wilkins, *28 co-authors* (incl. **A. R. Vernon**)
Laser-spectroscopy studies of the nuclear structure of neutron-rich radium
Phys. Rev. C, vol. 97, no. 2, pp. 19, 2018. doi:10.1103/PhysRevC.97.02430
T. Day Goodacre, *29 co-authors* (incl. **A. R. Vernon**)
Radium Ionization Scheme Development: The First Observed Autoionizing

**States
and Optical Pumping Effects in the Hot Cavity Environment.**
Spectrochim. Acta Part B, October. doi:10.1103/J.SAB.2018.10.002
S. G. Wilkins, K. M. Lynch, *17 co-authors* (incl. **A. R. Vernon**)
Quadrupole moment of ^{203}Fr
Phys. Rev. C, vol. 96, no. 3, p. 034317, 2017 doi:10.1103/PhysRevC.96.034317
R. P. De Groote, *22 co-authors* (incl. **A. R. Vernon**)
**Dipole and quadrupole moments $^{73-78}$ of Cu as a test of
the robustness of the $Z = 28$ shell closure near ^{78}Ni**
Phys. Rev. C, vol. 96, no. 4, pp. 16, 2017. doi:10.1103/PhysRevC.96.041302
2017   G. J. Farooq-Smith, **A. R. Vernon** et al.
**Probing the Ga_{31} ground-state properties in the region near $Z = 28$
with high-resolution laser spectroscopy,**
Phys. Rev. C, vol. 96, no. 4, p. 044324, Oct. 2017. doi:10.1103/PhysRevC.96.044324

Referees

Prof. Kieran Flanagan
The University of Manchester, Manchester, M13 9PL, United Kingdom
Email: kieran.flanagan-2@manchester.ac.uk   Phone: +44 161 275 4281
Prof. Ronald F. Garcia-Ruiz
Massachusetts Institute of Technology, Cambridge, MA 02139, USA
Email: rgarciar@mit.edu   Phone: +41 22 76 72563
Prof. Gerda Neyens
KU Leuven, Instituut voor Kern- en Stralingsfysica, B-3001 Leuven, Belgium
Email: gerda.neyens@fys.kuleuven.be   Phone: +41 22 76 75825